Simulated Annealing
and Boltzmann Machines

Wiley–Interscience Series in Discrete Mathematics and Optimization

Advisory Editors

Ronald L. Graham
AT & T Bell Laboratories, Murray Hill, New Jersey, USA

Jan Karel Lenstra
Centre for Mathematics and Computer Science, Amsterdam, The Netherlands

Robert E. Tarjan
Department of Computer Science, Princeton University, New Jersey, and AT & T Bell Laboratories, Murray Hill, New Jersey, USA

Simulated Annealing and Boltzmann Machines

A Stochastic Approach to Combinatorial Optimization
and Neural Computing

Emile Aarts

Philips Research Laboratories, Eindhoven
Eindhoven University of Technology, Eindhoven

Jan Korst

Philips Research Laboratories, Eindhoven

JOHN WILEY & SONS

Chichester · New York · Brisbane · Toronto · Singapore

Library of Congress Cataloging in Publication Data:

Aarts, E. H. L. (Emile H. L.)
    Simulated annealing and Boltzmann machines: a stochastic approach
to combinatorial optimization and neural computing / Emile Aarts,
Jan Korst.
        p.   cm.—(Wiley–Interscience series in discrete mathematics
and optimization)
    Bibliography: p.
    Includes index.
    ISBN 0 471 92146 7
    1. Simulated annealing (Mathematics) 2. Machine theory.
3. Neural computers. I. Korst, Jan. II. Title. III. Title:
Boltzmann machines. IV. Series.
    QA402.5.A22   1989
    519—dc19   88–20871
                                                                                 CIP

British Library Cataloguing in Publication Data:

Aarts, E. H. L. (Emile H. L.)
    Simulated annealing and Boltzmann
    machines: a stochastic approach to
    combinatorial optimization and neural
    computing.
    1. Mathematics. Combinatorial optimisation
    I. Title II. Korst, Jan
    511'.6
ISBN 0 471 92146 7

Printed and bound in Great Britain by the Anchor Press, Tiptree, Essex

Preface

Annealing is the physical process of heating up a solid until it melts, followed by cooling it down until it crystallizes into a state with a perfect lattice. During this process, the free energy of the solid is minimized. Practice shows that the cooling must be done carefully in order not to get trapped in locally optimal lattice structures with crystal imperfections.

In combinatorial optimization, we can define a similar process. This process can be formulated as the problem of finding - among a potentially very large number of solutions - a solution with minimal cost. Now, by establishing a correspondence between the cost function and the free energy, and between the solutions and the physical states, we can introduce a solution method in the field of combinatorial optimization based on a simulation of the physical annealing process. The resulting method is called *Simulated Annealing*.

Salient features of this method are its general applicability and its ability to obtain solutions arbitrarily close to an optimum. A major drawback however is that finding high-quality solutions may require large computational efforts.

A substantial reduction of the computational effort required by the simulated annealing algorithm may be achieved by using computational models based on massively parallel execution. An example of such a model is the *Boltzmann machine*.

The Boltzmann machine is a neural network model and belongs to the class of connectionist models. A Boltzmann machine consists of a large network of simple computing elements, called units, that are connected in some way. The units can have two states, either 'on' or 'off', and the connections have real-valued strengths that impose local constraints on the states of the individual units. A consensus function gives a quantitative measure for the 'goodness' of a global configuration of the Boltzmann machine, determined by the states of all individual units.

Massive parallelism and distributed representations are the salient features of Boltzmann machines. These features lead to a conceptually simple, yet powerful cooperative computational model, that can be viewed as an in-

v

teresting architectural blueprint for future parallel computers that are suitable for parallel execution of the simulated annealing algorithm. Furthermore, the model can cope with higher-order optimization problems such as learning. Its structure and its capability to learn by self-organisation justify a comparison with the human brain.

The aim of this book is to bring together in one volume the basic elements of the theory of simulated annealing and the model of Boltzmann machines. The material is presented in a step-by-step way, using the theory of combinatorial optimization as a guideline. The book is intended as a text for courses at the graduate level, but also for special courses at the undergraduate level. Furthermore, it may serve as a self-study guide for scientists that are professionally interested in general approximation methods for combinatorial optimization.

The book consists of two parts. In Part I the simulated annealing algorithm is treated in detail. Chapter 1 starts off with the general problem formulation of combinatorial optimization and a discussion of local search algorithms, based on the concept of neighbourhood structures. In Chapter 2 the analogy with the physical annealing process is used to introduce the simulated annealing algorithm. Furthermore, in this chapter, the gross features of the annealing algorithm are analysed using concepts from statistical physics such as equilibrium statistics and ensemble averages.

In Chapter 3 the asymptotic convergence of the simulated annealing algorithm is analysed for the standard choice of the generation and acceptance probabilities. The analysis is based on the theory of homogeneous and inhomogeneous Markov chains and the algorithm is rigorously proved to converge to globally optimal solutions. The analysis furthermore includes a discussion on the speed of convergence.

In practical implementations, the algorithm becomes an approximation algorithm since convergence to globally optimal solutions only holds asymptotically. In this case the convergence of the algorithm depends on a so-called cooling schedule, i.e. a set of parameters determining the efficiency and effectiveness of the algorithm. In Chapter 4 a polynomial-time cooling schedule is derived and its performance is analysed empirically by applying the simulated annealing algorithm with this cooling schedule to an instance of the travelling salesman problem.

To illustrate the practical use of the simulated annealing algorithm, we discuss in Chapter 5 the application of the algorithm to a number of combinatorial optimization problems. Furthermore, a review is given of a large number of combinatorial optimization problems from different fields to which the algorithm has been applied. The review is presented as a bibliography.

One of the conclusions reached in Chapter 5 is that the efficiency of the simulated annealing algorithm is poor for a number of problems. In Chapter 6 we discuss a number of possible approaches to speeding up the algorithm by execution on parallel computers. In addition to a discussion of the general aspects of designing parallel simulated annealing algorithms, we present three specific examples of parallel annealing algorithms together with an analysis of their performance.

In Part II, the problem of designing parallel annealing algorithms is reexamined on the basis of the model of Boltzmann machines. To introduce the model we discuss in Chapter 7 the subject of neural computing, which serves as a background for understanding the basic concepts of connectionist models and neural networks.

In Chapter 8 the model of Boltzmann machines is formally introduced by means of a graph-theoretical formulation and again the theory of Markov chains is used to model the state transitions in a Boltzmann machine.

Many combinatorial optimization problems can be formulated as instances of 0-1 programming problems. By assigning the states of a Boltzmann machine to the variables of a 0-1 programming problem and implementing the cost function and the constraints, that go with the problem, as connections with well-determined strengths, it can be shown, under certain weak conditions, that maximizing the consensus function of the Boltzmann machine becomes equivalent to solving the combinatorial optimization problem at hand. This approach is discussed in Chapter 9 for a number of well-known combinatorial optimization problems.

In Chapter 10 we discuss the subject of classification and Boltzmann machines. Classification problems, which play an important role in the field of pattern recognition, can be considered as a special class of combinatorial optimization problems, and can be solved accordingly with a Boltzmann machine. However, the cost function for these problems is usually specified implicitly; this as opposed to ordinary optimization problems for which the cost function is explicitly specified by an analytical expression.

For this class of problems, appropriate connection strengths can be obtained by learning. Learning in a Boltzmann machine is discussed in Chapter 11. The approach amounts to constructing an explicit cost function for problems that are implicitly specified by means of a number of learning examples. The construction of the explicit cost function is governed by a learning algorithm. The learning algorithm presented in Chapter 11 is based on the asymptotic convergence properties of the simulated annealing algorithm.

Learning in a Boltzmann machine leads to characteristic features such as memory and association. These features are illustrated by discussing a number of examples.

To conclude this preface, we would like to acknowledge the assistance of many colleagues in preparing this book. We are especially grateful for the help of Frans de Bont, Ronan Burgess, Peter van Laarhoven, Jan Karel Lenstra, Peter Schuur, and Patrick Zwietering.

Eindhoven, The Netherlands   Emile H.L. Aarts
June 1988   Jan H.M. Korst

Contents

PART I

Simulated Annealing

Simulated annealing

CHAPTER 1

Combinatorial Optimization

Solving a combinatorial optimization problem amounts to finding the 'best' or 'optimal' solution among a finite or countably infinite number of alternative solutions [Papadimitriou & Steiglitz, 1982]. In this book we concentrate on problems that can be formulated unambiguously in terms of mathematical terminology and notation. Furthermore, it is assumed that the quality of a solution is quantifiable and that it can be compared with that of any other solution. Finally, we assume the set of solutions to be finite.

Over the past few decades, a wide variety of such problems has emerged from such diverse areas as management science, computer science, engineering, VLSI design, etc. Among all combinatorial optimization problems the *travelling salesman problem* is probably the best known [Lawler, Lenstra, Rinnooy Kan & Shmoys, 1985]. In this problem a salesman, starting from his home city, is to visit each city on a prescribed list exactly once and then return home, in such a way that the length of his tour is minimal. The importance of the travelling salesman problem lies in the fact that it combines the typical features of a large class of combinatorial optimization problems and contains both the elements that attract mathematicians to particular problems: simplicity of statement and difficulty of solution [Garfinkel, 1985].

Considerable effort has been devoted to constructing and investigating methods for solving to optimality or proximity combinatorial optimization problems. Integer, linear and non-linear programming, as well as dynamic programming have seen major breakthroughs in recent years. A survey of the historical developments is given in the succession of monographs by Ford & Fulkerson [1962], Dantzig [1963], Simonnard [1966] Hu [1970], Aho, Hopcroft & Ullman [1974], Lawler [1976], Lawler, Lenstra, Rinnooy Kan & Shmoys [1985], and Schrijver [1986].

An important achievement in the field of combinatorial optimization, obtained in the late 1960's, is the conjecture - which is still unverified - that there exists a class of combinatorial optimization problems of such inherent complexity that any algorithm, solving each instance of such a problem to optimality, requires a computational effort that grows superpolynomially with the size of the problem. This conjecture resulted in a distinction between easy and hard problems. During the 1970's, developments in theoretical computer science have provided a rigorous formulation of this conjecture. The resulting theory of *NP-completeness* has greatly increased the insight in the relationship among hard problems [Miller & Thatcher, 1972]. The seminal papers within this respect are by Cook [1971; 1972], Karp [1972], and Levin [1973].

Over the years it has been shown that many theoretical and practical combinatorial optimization problems belong to the class of *NP-complete problems*. An impressive overview of problems in this class is given by Garey & Johnson [1979]. A direct consequence of the property of NP-completeness is that optimal solutions cannot be obtained in reasonable amounts of computation time.

However, large NP-complete problems still must be solved and in constructing appropriate algorithms one might choose between two options. Either one goes for optimality at the risk of very large, possibly impracticable, amounts of computation time, or one goes for quickly obtainable solutions at the risk of sub-optimality. The first option constitutes the class of *optimization algorithms*. Well-known examples are *enumeration methods* using cutting plane, branch and bound, or dynamic programming techniques [Papadimitriou & Steiglitz, 1982]. The second option constitutes the class of *approximation algorithms*, also often called *heuristic algorithms*; examples are *local search* and *randomization algorithms*. The division between the two classes is however not very strict. Some types of algorithm can be used for both purposes. For instance, by introducing heuristic bounding rules, a branch-and-bound algorithm can be easily transferred from an optimization to an approximation algorithm.

Furthermore, one may distinguish in both classes between general algorithms and tailored algorithms. *General algorithms* are applicable to a wide variety of problems and therefore may be called problem independent. *Tailored algorithms* use problem-specific information and their application is therefore limited to a restrictive set of problems. The intrinsic problem of tailored algorithms is that for each type of combinatorial optimization problem a new algorithm must be constructed that is tailored to that problem. Among the few exceptions are the edge-interchange algorithms for graph partitioning and travelling salesman problems, introduced by Kernighan & Lin [1970] and

Lin [1965], respectively. Of course, it is desirable to have an approximation algorithm of high quality, applicable to a wide variety of combinatorial optimization problems. Algorithms based on local search are typical examples of generally applicable approximation algorithms, but they are often of low quality. The *simulated annealing algorithm*, which is the main subject of the first part of this book, is a high quality general algorithm. In nature it is a randomization algorithm and, as we will show, it can be asymptotically viewed as an optimization algorithm. However, in any practical implementation it behaves as an approximation algorithm.

1.1 Combinatorial Optimization Problems

We now introduce a general formulation of combinatorial optimization problems.

Definition 1.1 A *combinatorial optimization problem* is either a *minimization problem* or a *maximization problem* and is specified by a set of problem instances. ∎

Definition 1.2 An *instance* of a combinatorial optimization problem can be formalized as a pair (S, f), where the *solution space* S denotes the finite set of all possible solutions and the *cost function* f is a mapping defined as

$$f : S \to \mathbb{R}. \tag{1.1}$$

In the case of minimization, the problem is to find a solution $i_{opt} \in S$ which satisfies

$$f(i_{opt}) \leq f(i), \quad \text{for all } i \in S. \tag{1.2}$$

In the case of maximization, i_{opt} satisfies

$$f(i_{opt}) \geq f(i), \quad \text{for all } i \in S. \tag{1.3}$$

Such a solution i_{opt} is called a *globally-optimal solution*, either *minimal* or *maximal*, or simply an *optimum*, either a *minimum* or a *maximum*; $f_{opt} = f(i_{opt})$ denotes the optimal cost, and S_{opt} the set of optimal solutions. ∎

In this book, unless explicitly stated otherwise, we consider combinatorial optimization problems as minimization problems. This can be done without loss of generality since maximization is equivalent to minimization after simply reversing the sign of the cost function.

The careful distinction between a problem and an instance of a problem is made so as to differentiate between one single 'input data', from which a solution can be obtained in an unambiguous way, and a set of input data which are generally generated in the same way.

Example 1.1 (Travelling salesman problem) We are given n cities and an $n \times n$-matrix $[d_{pq}]$, whose elements denote the distance between each pair p, q of the n cities. A *tour* is defined as a closed path visiting each city exactly once. The problem is to find a tour of minimal length. For this problem a solution is given by a cyclic permutation $\pi = (\pi(1), \ldots, \pi(n))$, where $\pi(k)$ denotes the successor city of city k, with $\pi^l(k) \neq k$, $l = 1, \ldots, n - 1$ and $\pi^n(k) = k$, for all k. Each solution then corresponds uniquely to a tour. The solution space is given by

$$S = \{\text{all cyclic permutations } \pi \text{ on } n \text{ cities}\}. \tag{1.4}$$

The cost function is defined as

$$f(\pi) = \sum_{i=1}^{n} d_{i, \pi(i)}, \tag{1.5}$$

i.e. $f(\pi)$ gives the length of the tour corresponding to π. Furthermore, we have $|S| = (n - 1)!$. ∎

1.2 Local Search

Local search algorithms constitute an interesting class of general approximation algorithms that are based on stepwise improvement on the value of the cost function by exploring neighbourhoods.[1] Furthermore, these algorithms have a strong relationship with the simulated annealing algorithm, and it is for this reason that we briefly discuss some properties of local search algorithms. The use of a local search algorithm presupposes the definition of solutions, a cost function and a neighbourhood structure.

Definition 1.3 Let (S, f) be an instance of a combinatorial optimization problem. Then a *neighbourhood structure* is a mapping

$$\mathcal{N} : S \to 2^S, \tag{1.6}$$

which defines for each solution $i \in S$ a set $S_i \subset S$ of solutions that are 'close' to i in some sense. The set S_i is called the *neighbourhood* of solution i, and

---

[1]Other algorithms in this class are *neighbourhood search* and *iterative improvement* algorithms.

each $j \in S_i$ is called a *neighbouring solution* or *neighbour* of i. Furthermore, we assume that $j \in S_i \Leftrightarrow i \in S_j$. ∎

Example 1.2 (*k*-change [Lin, 1965]) In the travelling salesman problem a neighbourhood structure $\mathcal{N}_k$, called the *k*-change, defines for each solution i a neighbourhood S_i consisting of the set of solutions that can be obtained from the given solution i by removing k edges from the tour corresponding to solution i and replacing them with k other edges such that again a tour is obtained. ∎

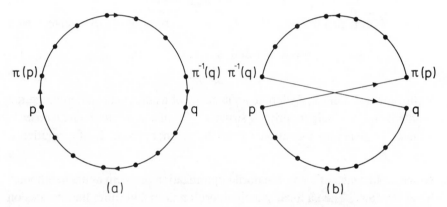

Figure 1.1: (a) A tour in an instance of the travelling salesman problem; (b) another tour obtained from (a) by an $N_2(p, q)$ change.

Example 1.3 (2-change [Lin, 1965]) Let (S, f) be an instance of a travelling salesman problem with n cities, and $\mathcal{N}_2$ be the 2-change neighbourhood structure. Then the 2-change $N_2(p, q)$ can be viewed as a reversal of the order in which the cities are traversed in between the cities p and q; see Figure 1.1. If the 2-change $N_2(p, q)$ changes a solution i into a solution j, then the cyclic permutations π_i and π_j are related in the following way. Let k denote the integer number for which $\pi_i^k(p) = q$, $1 < k < n$, then we have

$$\pi_j(p) = \pi_i^{-1}(q) = \pi_i^{k-1}(p),$$

$$\pi_j(\pi_i(p)) = q = \pi_i^k(p),$$

$$\pi_j(\pi_i^r(p)) = \pi_i^{r-1}(p), \qquad r = 2, \ldots, k-1,$$

$$\pi_j(s) = \pi_i(s), \qquad \text{otherwise.} \qquad (1.7)$$

Hence,

$$S_i = \{j \in S \mid \pi_j \text{ is obtained from } \pi_i \text{ by 2-change}\}, \qquad (1.8)$$

and

$$|S_i| \stackrel{\text{def}}{=} \Theta = (n-1)(n-2), \quad \text{for all } i. \qquad (1.9)$$

Furthermore, each solution j can be obtained from any other solution i by $n - 2$ 2-changes, i.e. for all $i, j \in S$ there exists a sequence $l_0, \ldots, l_{n-2} \in S$ such that

$$\pi_{l_0} = \pi_i \text{ and } \pi_{l_{n-2}} = \pi_j, \qquad (1.10)$$

and $\pi_{l_{k+1}}$ is obtained from π_{l_k}, $k = 0, \ldots, n-3$, by an $N_2(p, q)$ change, where

$$p = \pi_j^k(1) \text{ and } q = \pi_{l_k}(\pi_j^{k+1}(1)). \qquad (1.11)$$

■

Definition 1.4 Let (S, f) denote an instance of a combinatorial optimization problem and N a neighbourhood structure. Then a *generation mechanism* is a means of selecting a solution j from the neighbourhood S_i of a solution i.

■

Given an instance of a combinatorial optimization problem and a neighbourhood structure. Then a local search algorithm is an algorithm that iterates on a number of solutions. It starts off with a given initial solution, which is often randomly chosen. Next, by applying a generation mechanism, it continuously tries to find a better solution by searching the neighbourhood of the current solution for a solution with lower cost. If such a solution is found, the current solution is replaced by this solution. Otherwise the algorithm continues with the current solution. The algorithm terminates when no strict improvements can be obtained anymore. Figure 1.2 gives an example of a local search algorithm for minimization.

An important concept in the analysis of local search algorithms is that of local optimality, which can be defined as follows.

Definition 1.5 Let (S, f) be an instance of a combinatorial optimization problem and let N be a neighbourhood structure, then $\hat{\imath} \in S$ is called a *locally optimal solution* or simply a *local optimum* with respect to N if $\hat{\imath}$ is better than, or equal to, all its neighbouring solutions with regard to their cost. More specifically, in the case of minimization, $\hat{\imath}$ is called a *locally minimal solution* or simply a *local minimum* if

$$f(\hat{\imath}) \leq f(j), \quad \text{for all } j \in S_i, \qquad (1.12)$$

---

procedure LOCAL_SEARCH;

begin

   INITIALIZE (i_{start});

   $i := i_{start}$;

   repeat

      GENERATE (j from S_i);

      if $f(j) < f(i)$ **then** $i := j$;

   until $f(j) \geq f(i)$, for all $j \in S_i$;

end;

Figure 1.2: A pseudo PASCAL program of a local search algorithm.

and in the case of maximization, $\hat{\imath}$ is called a *locally maximal solution* or simply a *local maximum* if

$$f(\hat{\imath}) \geq f(j), \quad \text{for all } j \in S_{\hat{\imath}}. \tag{1.13}$$

∎

Definition 1.6 Let (S, f) be an instance of a combinatorial optimization problem and $\mathcal{N}$ a neighbourhood structure. Then $\mathcal{N}$ is called *exact* if, for each $\hat{\imath} \in S$ that is locally optimal with respect to $\mathcal{N}$, $\hat{\imath}$ is also globally optimal. ∎

Example 1.4 In the travelling salesman problem, $\mathcal{N}_2$ is not exact, but $\mathcal{N}_n$, where n denotes the number of cities, is exact. Moreover, in the latter case we have $S_i = S$, for all $i \in S$. ∎

Thus, by definition, local search algorithms terminate in a local optimum and, unless the neighbourhood structure is exact, there is generally no guarantee as to 'how far' this local optimum is from a global optimum [Johnson, Papadimitriou & Yannakakis, 1985]. However, exact neighbourhoods are in most cases impracticable since they lead to complete enumeration in a local search algorithm. This is for instance the case in the local search algorithm of Figure 1.2 when applied to the travelling salesman problem using the $\mathcal{N}_n$ neighbourhood structure. Thus, in local search algorithms, one generally resorts to non-exact neighbourhood structures resulting in locally optimal solutions.

   The quality of the local optimum obtained by a local search algorithm usually strongly depends on the initial solution and for most combinatorial optimization problems no guidelines are available for an appropriate choice of the initial solution. Furthermore, the *worst-case time complexity*, an upper bound on the computation time, is not known for many problems. For instance, the worst-case time complexity of Lin's well-known local search algorithm for the travelling salesman problem using 2-changes [Lin, 1965] is an open problem.[2]

It should be clear however that, in contrast to the disadvantages mentioned above, local search algorithms do have the advantage of being *generally applicable* and *flexible*: only the specification of solutions, a cost function and a neighbourhood structure are required, which are easy and straightforward for most problems, allowing different aspects to be handled.

To avoid some of the disadvantages mentioned above, whilst maintaining the basic principle of local search algorithms, i.e. iteration among neighbouring solutions, one might consider the following alternative approaches.

- Execution of the local search algorithm for a large number of initial solutions. Clearly, asymptotically - under the guarantee that all solutions have been used as initial solution - such an algorithm finds a global optimum with probability one.

- Introduction of a more complex neighbourhood structure, in order to be able to search a larger part of the solution space. Experience shows that finding such a neighbourhood structure is often cumbersome and requires expert knowledge of the combinatorial optimization problem at hand. It therefore usually turns the generally applicable local search algorithm into a tailored algorithm.

- Accepting in a limited way transitions corresponding to an increase in the value of the cost function; note that in local search algorithms only those transitions are accepted that correspond to a decrease in cost.

The simulated annealing algorithm closely follows the last alternative. The algorithm was independently introduced by Kirkpatrick, Gelatt and Vecchi [1982; 1983], and Černy [1985]. The name 'simulated annealing' originates from the analogy with the physical annealing process of solids and will be discussed in greater detail in the next chapter. Other names that have been used to denote the algorithm are *Monte Carlo annealing* [Jepsen & Gelatt,

---

[2]Recently, it was shown by Kern [1986a] that, for a limited class of problem instances, the average-case time complexity was given by a polynomial function in the size of the input.

1983], *probabilistic hill climbing* [Romeo & Sangiovanni-Vincentelli, 1985], *statistical cooling* [Aarts & Van Laarhoven, 1985a; Storer, Becker & Nicas, 1985], and *stochastic relaxation* [Geman & Geman, 1984].

A strong feature of the simulated annealing algorithm is that it finds high-quality solutions which do not strongly depend on the choice of the initial solution, i.e. the algorithm is *effective* and *robust*. Furthermore, it is possible to give a polynomial upper bound on the computation time for some implementations of the algorithm; see Chapter 4. Thus, the algorithm does not have the disadvantages of local search algorithms but it is as generally applicable, which makes it an interesting general approximation algorithm.

The remainder of Part I of this book is devoted to a detailed discussion of the simulated annealing algorithm and its application in the field of combinatorial optimization. In Chapter 2, the simulated annealing algorithm is introduced from the analogy with the physical annealing process. Furthermore, this analogy is used to derive a number of characteristic features of the algorithm. Next, in Chapter 3, a detailed treatment is given of the asymptotic convergence properties of the algorithm, based on the mathematical theory of Markov chains. Chapter 4 elaborates on the finite-time behaviour of the algorithm. Furthermore, an empirical study is presented based on numerical results, obtained by applying the algorithm to an instance of the travelling salesman problem. Chapter 5 discusses the application of the algorithm to a number of well-known combinatorial optimization problems. Chapter 6 concludes Part I with a discussion on parallel simulated annealing algorithms.

CHAPTER 2

Simulated Annealing

In the early 1980's Kirkpatrick, Gelatt & Vecchi [1982; 1983] and independently Černy [1985] introduced the concepts of annealing in combinatorial optimization. These concepts are based on a strong analogy between the physical annealing process of solids and the problem of solving large combinatorial optimization problems. In this chapter we will pursue this analogy in order to introduce the simulated annealing algorithm from an intuitive point of view.

2.1 The Metropolis Algorithm

In *condensed matter physics*, *annealing* is known as a thermal process for obtaining low energy states of a solid in a *heat bath*. The process contains the following two steps [Barker & Henderson, 1976; Kirkpatrick, Gelatt & Vecchi, 1982; 1983].

- Increase the temperature of the heat bath to a maximum value at which the solid melts.

- Decrease **carefully** the temperature of the heat bath until the particles arrange themselves in the ground state of the solid.

In the liquid phase all particles of the solid arrange themselves randomly. In the ground state the particles are arranged in a highly structured lattice and the energy of the system is minimal. The ground state of the solid is obtained only if the maximum temperature is sufficiently high and the cooling is done sufficiently slow. Otherwise the solid will be frozen into a meta-stable state rather than into the ground state.

The converse of annealing is a process known as *quenching* in which the temperature of the heat bath is instantaneously lowered. This again results in a meta-stable state.

The physical annealing process can be modelled successfully by using computer simulation methods from condensed matter physics. For a detailed overview see Binder [1978]. As far back as 1953, Metropolis, Rosenbluth, Rosenbluth, Teller & Teller [1953] introduced a simple algorithm for simulating the evolution of a solid in a heat bath to *thermal equilibrium*. The algorithm introduced by these authors is based on *Monte Carlo techniques* and generates a sequence of states of the solid in the following way. Given a current state i of the solid with *energy* E_i, then a subsequent state j is generated by applying a perturbation mechanism which transforms the current state into a next state by a small distortion, for instance by displacement of a particle. The energy of the next state is E_j. If the *energy difference*, $E_j - E_i$, is less than or equal to 0, the state j is accepted as the current state. If the energy difference is greater than 0, the state j is accepted with a certain probability which is given by

$$\exp\left(\frac{E_i - E_j}{k_B T}\right), \tag{2.1}$$

where T denotes the *temperature* of the heat bath and k_B a physical constant known as the *Boltzmann constant*. The acceptance rule described above is known as the *Metropolis criterion* and the algorithm that goes with it is known as the *Metropolis algorithm*.

If the lowering of the temperature is done sufficiently slow, the solid can reach thermal equilibrium at each temperature. In the Metropolis algorithm this is achieved by generating a large number of transitions at a given temperature value. Thermal equilibrium is characterized by the *Boltzmann distribution* [Toda, Kubo & Saitô, 1983]. This distribution gives the probability of the solid being in a state i with energy E_i at temperature T, and is given by

$$\mathbf{P}_T\{\mathbf{X} = i\} = \frac{1}{Z(T)} \exp\left(\frac{-E_i}{k_B T}\right), \tag{2.2}$$

where $\mathbf{X}$ is a stochastic variable denoting the current state of the solid. $Z(T)$ is the *partition function*, which is defined as

$$Z(T) = \sum_j \exp\left(\frac{-E_j}{k_B T}\right), \tag{2.3}$$

where the summation extends over all possible states. As we will see in the following sections, the Boltzmann distribution plays an essential role in the analysis of the simulated annealing algorithm.

2.2 The Simulated Annealing Algorithm

Returning to the simulated annealing algorithm, we can apply the Metropolis algorithm to generate a sequence of solutions of a combinatorial optimization problem. For this purpose we assume an analogy between a physical many-particle system and a combinatorial optimization problem based on the following equivalences.

- Solutions in a combinatorial optimization problem are equivalent to states of a physical system.

- The cost of a solution is equivalent to the energy of a state.

Next, we introduce a parameter which plays the role of the temperature. This parameter is called the *control parameter*.

The simulated annealing algorithm now can be viewed as an iteration of Metropolis algorithms, evaluated at decreasing values of the control parameter.

As in local search algorithms we assume the existence of a neighbour-hood structure and a generation mechanism. We now introduce the following definitions.

Definition 2.1 Let (S, f) denote an instance of a combinatorial optimization problem and i and j two solutions with cost $f(i)$ and $f(j)$, respectively. Then the *acceptance criterion* determines whether j is accepted from i by applying the following *acceptance probability*:

$$\mathbb{P}_c\{\text{accept } j\} = \begin{cases} 1 & \text{if } f(j) \leq f(i) \\ \exp\left(\frac{f(i)-f(j)}{c}\right) & \text{if } f(j) > f(i), \end{cases} \qquad (2.4)$$

where $c \in \mathbb{R}^+$ denotes the control parameter. ∎

Clearly, the generation mechanism corresponds to the perturbation mechanism in the Metropolis algorithm, whereas the acceptance criterion corresponds to the Metropolis criterion.

Definition 2.2 A *transition* is a combined action resulting in the transformation of a current solution into a subsequent one. The action consists of the following two steps: (i) application of the generation mechanism, (ii) application of the acceptance criterion. ∎

Let c_k denote the value of the control parameter and L_k the number of transitions generated at the kth iteration of the Metropolis algorithm. Then the simulated annealing algorithm can be described in pseudo PASCAL as is shown in Figure 2.1.

---

procedure SIMULATED_ ANNEALING;

begin

    INITIALIZE (i_{start}, c_0, L_0);

    $k := 0$;

    $i := i_{start}$;

    repeat

        for $l := 1$ **to** L_k **do**

        begin

            GENERATE $(j$ from $S_i)$;

            if $f(j) \leq f(i)$ **then** $i := j$

            else

            if $\exp\left(\frac{f(i)-f(j)}{c_k}\right) > \text{random}[0, 1)$ **then** $i := j$

        end;

        $k := k + 1$;

        CALCULATE_ LENGTH (L_k);

        CALCULATE_ CONTROL (c_k);

    until stopcriterion

end;

Figure 2.1: A pseudo PASCAL program of the simulated annealing algorithm.

A typical feature of the simulated annealing algorithms is that, besides accepting improvements in cost, it also to a limited extent accepts deteriorations in cost. Initially, at large values of c, large deteriorations will be accepted; as c decreases, only smaller deteriorations will be accepted and finally, as the value of c approaches 0, no deteriorations will be accepted at all. This feature means that the simulated annealing algorithm, in contrast to local search algorithms, can escape from local minima while it still exhibits the favourable features of local search algorithms, i.e. simplicity and general applicability.

Note that the probability of accepting deteriorations is implemented by comparing the value of $\exp((f(i) - f(j))/c)$ with a random number generated from a uniform distribution on the interval [0,1). Furthermore, it should be obvious that the speed of convergence of the algorithm is determined by the choice of the parameters L_k and c_k, $k = 0, 1, \ldots$. In Chapter 3 we will derive values for the parameters that assure convergence in distribution to the set of globally optimal solutions. In Chapter 4 more practical, implementation-oriented choices of the parameter values are discussed, leading to finite-time execution of the algorithm.

Comparing simulated annealing to local search it is evident that simulated annealing can be viewed as a generalization of local search. It becomes identical to some form of local search in the case where the value of the control parameter is taken equal to zero. In the analogy with physics, local search can be viewed as the aforementioned quenching process. With respect to a comparison between the performance of both algorithms we already mention that for most problems simulated annealing performs better than local search, repeated for a number of different initial solutions. We will return to this subject in Chapter 5.

2.3 Equilibrium Statistics

In this section we discuss a number of quantities that emerge from the theory of *statistical physics*. Using these quantities as a guideline, a number of characteristic features of the simulated annealing algorithm can be analysed. The discussion focusses on global aspects such as entropy and ensemble averages.

Starting off from the fundamental assumption of statistical physics, i.e. the *ergodicity hypothesis* [Toda, Kubo & Saitô, 1983], which states that a physical many-particle system is compatible with a statistical ensemble, and that the corresponding ensemble averages determine the averages of *observables* of the physical system, then a number of useful quantities can be derived for the physical system at thermal equilibrium. Examples of these quantities are the average energy, the spreading of the energy, and the entropy. Furthermore, as was first mentioned by Gibbs [1902], if the ensemble is stationary, which is the case if thermal equilibrium is achieved, its density is a function of the energy of the system. Moreover, applying the principle of equal probability, it can be shown that at thermal equilibrium the probability that the system is in a state i with energy E_i is given by the Gibbs or Boltzmann distribution of (2.2). This distribution allows an analytic evaluation of the quantities mentioned earlier.

We can now extend the analogy between statistical physics and optimization by simulated annealing in the following way.

Conjecture 2.1 *Given an instance* (S, f) *of a combinatorial optimization problem and a suitable neighbourhood structure then, after a sufficiently large number of transitions at a fixed value of* c, *applying the acceptance probability of (2.4), the simulated annealing algorithm will find a solution* $i \in S$ *with a probability equal to*

$$\mathbb{P}_c\{\mathbf{X} = i\} \overset{\text{def}}{=} q_i(c) = \frac{1}{N_0(c)} \exp\left(-\frac{f(i)}{c}\right), \tag{2.5}$$

where $\mathbf{X}$ *is a stochastic variable denoting the current solution obtained by the simulated annealing algorithm and*

$$N_0(c) = \sum_{j \in S} \exp\left(-\frac{f(j)}{c}\right) \tag{2.6}$$

denotes a normalization constant.   ■

What is meant in the conjecture by a 'suitable' neighbourhood structure is explained in greater detail in Chapter 3. The probability distribution of (2.5) is called the *stationary* or *equilibrium distribution* and is the equivalent of the Boltzmann distribution of (2.2). The normalization constant $N_0(c)$ is the equivalent of the partition function of (2.3).

Corollary 2.1 *Given an instance* (S, f) *of a combinatorial optimization problem and a suitable neighbourhood structure. Furthermore, let the stationary distribution be given by (2.5), then*

$$\lim_{c \downarrow 0} q_i(c) \overset{\text{def}}{=} q_i^* = \frac{1}{|S_{opt}|} \chi_{(S_{opt})}(i), ^{1} \tag{2.7}$$

where S_{opt} *denotes the set of globally optimal solutions.*

Proof Using the fact that for all $a \leq 0, \lim_{x \downarrow 0} e^{\frac{a}{x}} = 1$ if $a = 0$, and 0 otherwise, we obtain

$$\lim_{c \downarrow 0} q_i(c) = \lim_{c \downarrow 0} \frac{\exp\left(-\frac{f(i)}{c}\right)}{\sum_{j \in S} \exp\left(-\frac{f(j)}{c}\right)}$$

$$= \lim_{c \downarrow 0} \frac{\exp\left(\frac{f_{opt} - f(i)}{c}\right)}{\sum_{j \in S} \exp\left(\frac{f_{opt} - f(j)}{c}\right)}$$

---

[1]Let A and $A' \subset A$ be two sets. Then the characteristic function $\chi_{(A')} : A \to \{0, 1\}$ of the set A' is defined as $\chi_{(A')}(a) = 1$ if $a \in A'$, and $\chi_{(A')}(a) = 0$ otherwise.

$$= \lim_{c \downarrow 0} \frac{1}{\sum_{j \in S} \exp\left(\frac{f_{opt} - f(j)}{c}\right)} \chi_{(S_{opt})}(i)$$

$$+ \lim_{c \downarrow 0} \frac{\exp\left(\frac{f_{opt} - f(i)}{c}\right)}{\sum_{j \in S} \exp\left(\frac{f_{opt} - f(j)}{c}\right)} \chi_{(S \setminus S_{opt})}(i)$$

$$= \frac{1}{|S_{opt}|} \chi_{(S_{opt})}(i) + 0,$$

which completes the proof. ∎

The result of this corollary is very important since it guarantees *asymptotic convergence* of the simulated annealing algorithm to the set of globally optimal solutions under the condition that the stationary distribution of (2.5) is attained at each value of c. We return to this property in Chapter 3.

By using the stationary distribution of (2.5), a set of useful quantities can be defined for optimization problems in a way similar to that for physical many-particle systems [Toda, Kubo & Saitô, 1983] or statistical ensembles [Feller, 1950]. Here we define the following ones.

Definition 2.3 The *expected cost* $\mathbb{E}_c(f)$ at equilibrium is defined as

$$\mathbb{E}_c(f) \overset{\text{def}}{=} \langle f \rangle_c$$

$$= \sum_{i \in S} f(i) \mathbb{P}_c \{ \mathbf{X} = i \}$$

$$= \sum_{i \in S} f(i) q_i(c). \tag{2.8}$$

∎

Definition 2.4 The *expected square cost* $\mathbb{E}_c(f^2)$ at equilibrium is defined as

$$\mathbb{E}_c(f^2) \overset{\text{def}}{=} \langle f^2 \rangle_c$$

$$= \sum_{i \in S} f^2(i) \mathbb{P}_c \{ \mathbf{X} = i \}$$

$$= \sum_{i \in S} f^2(i) q_i(c). \tag{2.9}$$

∎

Definition 2.5 The *variance* $\text{Var}_c(f)$ of the cost at equilibrium is defined as

$$\text{Var}_c(f) \overset{\text{def}}{=} \sigma_c^2$$

$$= \sum_{i \in S} (f(i) - \mathbb{E}_c(f))^2 \mathbb{P}_c\{\mathbf{X} = i\}$$

$$= \sum_{i \in S} (f(i) - \langle f \rangle_c)^2 q_i(c)$$

$$= \langle f^2 \rangle_c - \langle f \rangle_c^2. \qquad (2.10)$$

■

The notations $\langle f \rangle_c$, $\langle f^2 \rangle_c$ and σ_c^2 are introduced as shorthand notations, and will be used throughout the remainder of this book.

Definition 2.6 The *entropy* at equilibrium is defined as

$$S_c = - \sum_{i \in S} q_i(c) \ln q_i(c). \qquad (2.11)$$

■

The entropy is a natural measure of the amount of disorder or information in a 'system'. It is an important quantity to which we return later in greater detail.

Corollary 2.2 *Let the stationary distribution be given by (2.5), then the following relations hold:*

$$\frac{\partial}{\partial c} \langle f \rangle_c = \frac{\sigma_c^2}{c^2} \qquad (2.12)$$

and

$$\frac{\partial}{\partial c} S_c = \frac{\sigma_c^2}{c^3}. \qquad (2.13)$$

Proof The relations can be straightforwardly verified by using the definitions of (2.8)-(2.11) and substituting the expression for the stationary distribution given by (2.5). ■

The expressions (2.8)-(2.13) are well known in statistical physics and play an important role in the analysis of the mechanics of large physical ensembles at thermal equilibrium. Below, we elaborate in some more detail on a number of these quantities and discuss some aspects that are of interest for the analysis of the simulated annealing algorithm.

Corollary 2.3 *Let the stationary distribution be given by (2.5). Then we have*

$$\lim_{c \to \infty} \langle f \rangle_c \overset{\text{def}}{=} \langle f \rangle_\infty = \frac{1}{|S|} \sum_{i \in S} f(i), \tag{2.14}$$

$$\lim_{c \downarrow 0} \langle f \rangle_c = f_{opt}, \tag{2.15}$$

$$\lim_{c \to \infty} \sigma_c^2 \overset{\text{def}}{=} \sigma_\infty^2 = \frac{1}{|S|} \sum_{i \in S} (f(i) - \langle f \rangle_\infty)^2, \tag{2.16}$$

$$\lim_{c \downarrow 0} \sigma_c^2 = 0, \tag{2.17}$$

$$\lim_{c \to \infty} S_c \overset{\text{def}}{=} S_\infty = \ln |S|, \tag{2.18}$$

and

$$\lim_{c \downarrow 0} S_c \overset{\text{def}}{=} S_0 = \ln |S_{opt}|. \tag{2.19}$$

Proof The relations can be easily verified by using the definitions of the expected cost (2.8), the variance (2.10) and the entropy (2.11), and by substituting the stationary distribution of (2.5) and applying similar arguments as in the proof of Corollary 2.1. ∎

We remark that in physics, assuming that there is only one ground state, (2.19) reduces to

$$S_0 = \ln(1) = 0, \tag{2.20}$$

which is known as the *third law of thermodynamics*. Furthermore, we remark that the entropy can be interpreted as a natural measure of the order of a physical system; high entropy values correspond to chaos; low entropy values to order [Toda, Kubo & Saitô, 1983]. A similar definition of the entropy as given in (2.11) is known in information theory, where it is viewed as a quantitative measure of the information content of a system [Shannon, 1948]. In the case of simulated annealing, or optimization in general, the entropy can be interpreted as a quantitative measure for the degree of optimality. This is illustrated in the next section.

Using (2.12) and (2.13), it follows that during execution of the simulated annealing algorithm the expected cost and entropy decrease monotonically - provided equilibrium is reached at each value of the control parameter - to their final values, i.e. f_{opt}, and $\ln |S_{opt}|$, respectively.

The dependence of the stationary distribution of (2.5) on the control parameter c is the subject of the following corollary.

Corollary 2.4 *Let* (S, f) *denote an instance of a combinatorial optimization problem with* $S_{opt} \neq S$, *and let* $q_i(c)$ *denote the stationary distribution associated with the simulated annealing algorithm and given by (2.5). Then we have*

(*i*) $\forall i \in S_{opt}$:

$$\frac{\partial}{\partial c} q_i(c) < 0, \tag{2.21}$$

(*ii*) $\forall i \notin S_{opt}, f(i) \geq \langle f \rangle_\infty$:

$$\frac{\partial}{\partial c} q_i(c) > 0, \tag{2.22}$$

(*iii*) $\forall i \notin S_{opt}, f(i) < \langle f \rangle_\infty, \exists \tilde{c}_i > 0$:

$$\frac{\partial}{\partial c} q_i(c) \quad > \quad 0 \text{ if } c < \tilde{c}_i$$
$$= \quad 0 \text{ if } c = \tilde{c}_i$$
$$< \quad 0 \text{ if } c > \tilde{c}_i. \tag{2.23}$$

Proof From (2.6) we can derive the following expression:

$$\frac{\partial}{\partial c} N_0(c) = \sum_{j \in S} \frac{f(j)}{c^2} \exp\left(\frac{-f(j)}{c}\right).$$

Hence, we obtain

$$\frac{\partial}{\partial c} q_i(c) \quad = \quad \frac{\partial}{\partial c} \frac{\exp\left(\frac{-f(i)}{c}\right)}{N_0(c)}$$

$$= \quad \left\{ \frac{f(i)}{c^2} \frac{\exp\left(\frac{-f(i)}{c}\right)}{N_0(c)} - \frac{\exp\left(\frac{-f(i)}{c}\right)}{N_0^2(c)} \frac{\partial}{\partial c} N_0(c) \right\}.$$

$$= \quad \frac{q_i(c)}{c^2} f(i) - \frac{q_i(c)}{c^2} \frac{\sum_{j \in S} f(j) \exp\left(\frac{-f(j)}{c}\right)}{N_0(c)}$$

$$= \quad \frac{q_i(c)}{c^2} (f(i) - \langle f \rangle_c). \tag{2.24}$$

Thus, the sign of $\frac{\partial}{\partial c} q_i(c)$ is determined by the sign of $f(i) - \langle f \rangle_c$ since $\frac{q_i(c)}{c^2} > 0$, for all $i \in S$ and $c > 0$.

From (2.12), (2.14), and (2.15) we have that $\langle f \rangle_c$ increases monotonically from f_{opt} to $\langle f \rangle_\infty$ with increasing c, provided $S_{opt} \neq S$. The proof of the theorem is now straightforward.

If $i \in S_{opt}$ and $S \neq S_{opt}$, then $f(i) < \langle f \rangle_c$. Hence, $\frac{\partial}{\partial c} q_i(c) < 0$ (cf. (2.24)), which completes the proof of part (i).

If $i \notin S_{opt}$, then the sign of $\frac{\partial}{\partial c} q_i(c)$ depends on the value of $\langle f \rangle_c$. Hence, using (2.24), we have that $\forall i \in S \backslash S_{opt}$: $\frac{\partial}{\partial c} q_i(c) > 0$ if $f(i) \geq \langle f \rangle_\infty$, whereas $\forall i \in S \backslash S_{opt}$, where $f(i) < \langle f \rangle_\infty$, there exists a $\tilde{c}_i > 0$ at which $f(i) - \langle f \rangle_c$ changes sign. Consequently, we have

$$\frac{\partial}{\partial c} q_i(c) \quad > \quad 0 \ \text{ if } c < \tilde{c}_i$$
$$= \quad 0 \ \text{ if } c = \tilde{c}_i$$
$$< \quad 0 \ \text{ if } c > \tilde{c}_i.$$

This completes the proofs of parts (ii) and (iii). ∎

From Corollary 2.4 it follows that the probability of finding an optimal solution increases monotonically with decreasing c. Furthermore, for each solution, not being an optimal one, there exists a positive value of the control parameter $\tilde{c}_i$, such that for $c < \tilde{c}_i$, the probability of finding that solution decreases monotonically with decreasing c.

2.4 Characteristic Features

To study the characteristic features of the simulated annealing algorithm we implemented the algorithm for the EUR100 instance of the travelling salesman problem. The EUR100 is a symmetric, Euclidean instance of the travelling salesman problem defined on a set of 100 large European cities. A description of this problem instance is given in the Appendix. The minimum tour length in this problem instance is 21,134 [Aarts, Korst & Van Laarhoven, 1988] and the corresponding tour is shown in the Appendix. Implementation of the simulated annealing algorithm was done in PASCAL on a VAX 11/785 according to the procedure SIMULATED_ ANNEALING presented in Section 2.1, using an $\mathcal{N}_2$ neighbourhood structure; see Example 1.3.

Before we continue our discussion we need the following definition of the acceptance ratio.

Definition 2.7 The *acceptance ratio* $\chi(c)$ in the simulated annealing algorithm is defined as

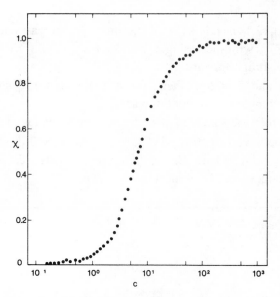

Figure 2.2: Acceptance ratio as a function of the control parameter c for the EUR100 problem instance.

$$\chi(c) = \left. \frac{\text{number of accepted transitions}}{\text{number of proposed transitions}} \right|_c . \qquad (2.25)$$

∎

Figure 2.2 shows the acceptance ratio for the EUR100 problem instance. The figure illustrates the behaviour as it would be expected from the acceptance criterion given in (2.4). At large values of c, virtually all proposed transitions are accepted. As c decreases, ever fewer proposed transitions are accepted, and finally, at very small values of c, no proposed transitions are accepted at all.

Figure 2.3 shows four solutions in the evolution of the optimization process carried out by simulated annealing for the EUR100 problem instance. The initial solution, shown in Figure 2.3a, is given by a random sequence of the 100 cities, which is far from an optimal solution. The solution looks very chaotic; the entropy is large and the corresponding value of the tour length l equals 129,965. In the course of the optimization process the solutions become less and less chaotic, see Figure 2.3b and Figure 2.3c, and the tour length is decreased to l=68,153 and l=33,048, respectively. Finally, the near-optimal solution shown in Figure 2.3d is obtained. This solution has a highly regular pattern for which the entropy is small and the tour length equals l=21,456.

Figure 2.4 shows (a) the normalized *average cost* and (b) the normalized

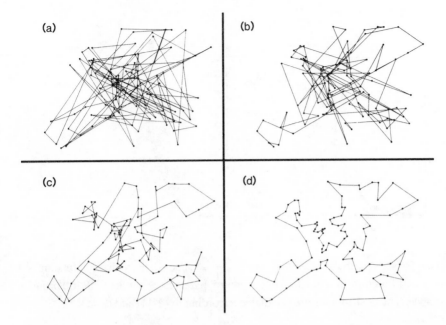

Figure 2.3: Solutions of the EUR100 problem instance at four different stages of the simulated annealing algorithm. (a) The initial tour (c=17.85) looks very chaotic and the entropy is large; (b) and (c) as the value of the control parameter decreases the tours become less chaotic (c=4.46 and 1.28, respectively) and the entropy decreases; (d) the final tour (c=0.06) is highly regular which is reflected by a small cost and entropy.

spreading of the cost as a function of the control parameter c. The data are obtained by applying the simulated annealing algorithm to the EUR100 problem instance, and the numerical values are calculated from the following expressions:

$$\bar{f}(c) = \frac{1}{L} \sum_{i=1}^{L} f_i(c) \qquad (2.26)$$

and

$$\sigma(c) = \left(\frac{1}{L} \sum_{i=1}^{L} (f_i(c) - \bar{f}(c))^2 \right)^{\frac{1}{2}}, \qquad (2.27)$$

where the average is taken over the values of the cost function $f_i(c)$, $i = 1, \ldots, L$, of L solutions generated at a given value of the control parameter c. Clearly, the average cost and spreading are approximations of the expected cost and the square root of the variance of the cost, respectively.

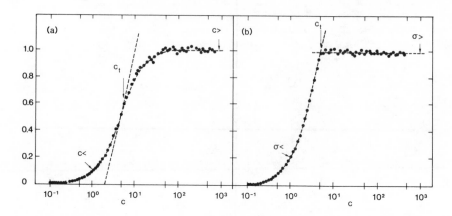

Figure 2.4: (a) Normalized average value $\frac{\overline{f}(c)-f_{opt}}{\langle f\rangle_\infty - f_{opt}}$, and (b) normalized spreading $\frac{\sigma(c)}{\sigma_\infty}$ of the cost function, as a function of the control parameter c for the EUR100 problem instance. The dashed lines are calculated according to (2.44) and (2.45).

The typical behaviour shown in Figure 2.4 is observed for many different problem instances and is reported in the literature by a number of authors [Aarts, Korst & Van Laarhoven, 1988; Hajek, 1985; Kirkpatrick, Gelatt & Vecchi, 1983; Van Laarhoven & Aarts, 1987; White, 1984].

From the figures we can deduce some characteristic features of the expected cost $\langle f\rangle_c$ and the variance σ_c^2 of the cost.

Firstly, it is observed that for large values of c the average and the spreading of the cost are about constant and equal to $\langle f\rangle_\infty$ and σ_∞, respectively. This behaviour is directly explained from (2.14) and (2.16), from which it is predicted that both the average value and the spreading of the cost function are constant at large c-values.

Secondly, we observe that there exists a threshold value c_t of the control parameter for which

$$\langle f\rangle_{c_t} \approx \frac{1}{2}(\langle f\rangle_\infty + f_{opt}) \tag{2.28}$$

and

$$\sigma_c^2 \quad \approx \quad \sigma_\infty^2 \quad \text{if } c \geq c_t$$
$$< \quad \sigma_\infty^2 \quad \text{if } c < c_t. \tag{2.29}$$

Moreover, we mention that c_t is roughly that value of c for which $\chi(c) \approx 0.5$, cf. Figures 2.2 and 2.4.

In the next section we show how the typical behaviour observed in Figures 2.4a and 2.4b can be modelled using simple statistical concepts such as the solution density and the stationary distribution.

2.5 A Quantitative Analysis

To model the characteristic features of the simulated annealing algorithm observed in the previous section, we discuss an analytical approach to calculate the expected cost $\langle f \rangle_c$ and the variance σ_c^2 of the cost.

Definition 2.8 Let (S, f) be an instance of a combinatorial optimization problem. Then the *solution density* $w(f)$ is defined by

$$w(f)df = \frac{1}{|S|}|\{i \in S | f \le f(i) < f + df\}|. \tag{2.30}$$

∎

In the case of the simulated annealing algorithm the equilibrium distribution of (2.5) can then be rewritten as

$$\Omega(f, c)df = \frac{w(f) \exp\left(\frac{f_{opt} - f}{c}\right) df}{\int_{f_{min}}^{f_{max}} w(f') \exp\left(\frac{f_{opt} - f'}{c}\right) df'}, \tag{2.31}$$

where f_{min} and f_{max} denote the minimum and maximum values of the cost function, respectively. Clearly, we have $f_{min} = f_{opt}$. Furthermore, we have

$$\langle f \rangle_c = \int_{f_{min}}^{f_{max}} f' \Omega(f', c)df' \tag{2.32}$$

and

$$\sigma_c^2 = \int_{f_{min}}^{f_{max}} ((f' - \langle f \rangle_c)^2 \Omega(f', c)df'. \tag{2.33}$$

Given an analytical expression for the solution density $w(f)$, it is possible to evaluate the integrals of (2.32) and (2.33). To estimate $w(f)$ for a particular combinatorial optimization problem is in most cases very hard. Indeed, $w(f)$ may vary drastically for different specific problem instances, especially for f-values close to f_{opt}. However, we can give some global estimates based on the following arguments. The threshold value c_t marks the transition of the observed solutions from one region to another. We argue that these regions correspond to different (dominant) contributions of the solution density $w(f)$ to the *observed solution density* $\Omega(f, c)$. This leads to the following postulate.

Postulate 2.1 *Let R_1 and R_2 be two regions of the value of the cost function, where R_1 denotes the region of a few standard deviations σ_∞ around $\langle f \rangle_\infty$ and R_2 the region close to f_{min}. Then, for a typical combinatorial optimization problem, $\omega(f)$ is given by a normal distribution $\omega_N(f)$ in the region R_1 and by an exponential distribution $\omega_E(f)$ in the region R_2. Furthermore, we conjecture that the number of solutions in R_1 is much larger than the number of solutions in R_2.*   ∎

From the postulate and (2.31) it follows that, for large values of c, $\Omega(f, c)$ is dominated by $\omega_N(f)$, whereas, for small values of c, the contribution of $\omega_E(f)$ is dominant. Thus, by splitting the contributions in the two regions, we can now evaluate the integerals of (2.32) and (2.33).

Region R_1 Writing $\omega_N(f)$ as

$$\omega_N(f) \propto \exp\left(-\frac{(f - \langle f \rangle_\infty)^2}{2\sigma_\infty^2}\right), \tag{2.34}$$

we obtain for $c \in R_1$

$$\Omega(f, c) \approx \frac{\exp\left(-\frac{(f - (\langle f \rangle_\infty - \frac{\sigma_\infty^2}{c}))^2}{2\sigma_\infty^2}\right)}{\int_{f_{min}}^{f_{max}} \exp\left(-\frac{(f' - (\langle f \rangle_\infty - \frac{\sigma_\infty^2}{c}))^2}{2\sigma_\infty^2}\right) dC'}$$

$$= \frac{1}{N(c)} \exp\left(-\frac{(f - (\langle f \rangle_\infty - \frac{\sigma_\infty^2}{c}))^2}{2\sigma_\infty^2}\right), \tag{2.35}$$

where f_{max} denotes the maximum value of the cost function and

$$N(c) = \frac{1}{2}\sigma_\infty\sqrt{2\pi} \left\{ \mathrm{erf}\left(\frac{f_{max} - (\langle f \rangle_\infty + \frac{\sigma_\infty^2}{c})}{\sigma_\infty\sqrt{2}}\right) \right.$$

$$\left. -\mathrm{erf}\left(\frac{f_{min} - (\langle f \rangle_\infty + \frac{\sigma_\infty^2}{c})}{\sigma_\infty\sqrt{2}}\right) \right\}. \tag{2.36}$$

Consequently, since the expression given by (2.35) again corresponds to a normal distribution, one obtains

$$\langle f \rangle_c \approx \langle f \rangle_\infty - \frac{\sigma_\infty^2}{c} \tag{2.37}$$

and

$$\sigma_c^2 \approx \sigma_\infty^2. \tag{2.38}$$

Region R_2 The solution density $\omega_{\mathcal{E}}(f)$ is given by an exponential distribution of the form

$$\omega_{\mathcal{E}}(f) \propto \exp((f - f_{min})\gamma), \tag{2.39}$$

for some constant γ $(0 < \gamma < c^{-1})$. Thus, for $c \in R_2$ we obtain

$$
\begin{aligned}
\Omega(f, c) &\approx \frac{\exp\left\{(f_{min} - f)\left(\frac{1-\gamma c}{c}\right)\right\}}{\int_{f_{min}}^{f_{max}} \exp\left\{(f_{min} - f')\left(\frac{1-\gamma c}{c}\right)\right\} df'} \\
&= \frac{1}{M(c)}\left(\frac{1-\gamma c}{c}\right) \exp\left\{(f_{min} - f)\left(\frac{1-\gamma c}{c}\right)\right\}, \tag{2.40}
\end{aligned}
$$

where

$$M(c) = 1 - \exp\left\{(f_{min} - f_{max})\left(\frac{1-\gamma c}{c}\right)\right\}. \tag{2.41}$$

Consequently, assuming that $f_{max} \gg f_{min}$, we obtain

$$\langle f \rangle_c - f_{min} \propto \frac{c}{1 - \gamma c} = c(1 + \gamma c + (\gamma c)^2 + \cdots), \tag{2.42}$$

and

$$\sigma_c^2 \propto \left(\frac{c}{1 - \gamma c}\right)^2 = c^2(1 + 2\gamma c + 3(\gamma c)^2 + \cdots). \tag{2.43}$$

From these derivations it follows that

- for large values of c, $\langle f \rangle_c$ is linear in c^{-1} and σ_c^2 is constant, whereas

- for small values of c, $\langle f \rangle_c$ is proportional to c and σ_c^2 to c^2.

Using the threshold value c_t as the value separating the two regions R_1 and R_2, then the results obtained above suggest a simple parametrization of the following form:

$$
\langle f \rangle_c =
\begin{cases}
f_< = f_{min} + N_t \left(\langle f \rangle_\infty - f_{min} - \frac{\sigma_\infty^2}{c_t}\right)\left(\frac{c}{1-\gamma c}\right) & \text{if } c \le c_t \\
f_> = \langle f \rangle_\infty - \frac{\sigma_\infty^2}{c} & \text{if } c > c_t,
\end{cases}
\tag{2.44}
$$

and

$$
\sigma_c^2 =
\begin{cases}
\sigma_<^2 = N_t^2 \sigma_\infty^2 \left(\frac{c}{1-\gamma c}\right)^2 & \text{if } c \le c_t \\
\sigma_>^2 = \sigma_\infty^2 & \text{if } c > c_t,
\end{cases}
\tag{2.45}
$$

where

$$N_t = \frac{1 - \gamma c_t}{c_t}, \tag{2.46}$$

and

$$c_t = \frac{2\sigma_\infty^2}{\langle f \rangle_\infty - f_{min}}. \tag{2.47}$$

The expression of (2.47) is obtained from (2.28) and (2.44).

The curves corresponding to the parametrization of (2.44)-(2.47) applied to the EUR100 problem instance are drawn in Figure 2.4 with $\gamma = 0.01$ and $c_t = 4.9$. The value of c_t was obtained from (2.47) using $\sigma_\infty = 5,300$ and $\langle f \rangle_\infty = 135,300$. The value of γ was determined experimentally. The curves fit the data very well; the correlation coefficient $\chi^2 = 0.91$. The parametrization formulas of (2.44)-(2.47) have also been applied successfully to a number of other problem instances and we conclude that they represent accurately the characteristic features of the expected cost and the variance of the cost, obtained by the simulated annealing algorithm when applied to these problem instances.

We end this section with some remarks. Evidently, the structure of the solution space differs from one problem instance to another and the distributions used in both regions are approximations. The normal distribution $\omega_N(f)$ is also reported by other authors [Hajek, 1985; White, 1984]. It corresponds to a distribution that maximizes the entropy, subject to the observed values of $\langle f \rangle_\infty$ and σ_∞ [Aarts, Korst & Van Laarhoven, 1988], and can be explained as follows. If the number of solutions is extremely large and the values of the cost function are distributed sufficiently uniformly (no clustering), then the number of degrees of freedom is large and the structure only plays a minor role. Consequently, the solution density is approximately given by a normal distribution which is typical of disordered systems [Binder, 1978]. In the region close to f_{min} the structure of a combinatorial optimization problem is important and strongly determines the solution density $\omega(f)$. Therefore, the exponential distribution $\omega_\varepsilon(f)$ given in (2.39) should be treated with caution. It can be argued that it holds for a number of examples [Hajek, 1985; Palmer, 1985]. Furthermore, as was shown above, it reproduces the behaviour of the simulated annealing algorithm very well.

Finally, we mention that the results of the quantitative analysis presented above are strongly based on the choices for the solution densities which, as was argued above, can only be obtained from rather intuitive arguments. In a more careful analysis, Van Laarhoven [1988] uses a *Bayesian approach* to estimate the solution densities from statistical experiments. In his approach,

Van Laarhoven views the execution of the Metropolis algorithm as a random experiment, with parameters that determine its outcome. The parameters he uses are the frequencies of occurence of the values of the cost function, and the minimal value of the cost function. Assuming a given *prior distribution* on the values of these parameters, and using the outcomes of an experiment given by the values of the cost function observed during the execution of the Metropolis algorithm, Bayes' theorem is used to derive a *posterior distribution* on the values of the parameters. Clearly, this approach does not use any assumption on the solution densities but estimates the densities from the frequencies of the solutions observed during execution of the Metropolis algorithm. This approach is used by Van Laarhoven to calculate (i) the expected cost at each value of the control parameter, and (ii) the expected minimum value of the cost function. Numerical experiments carried out for the travelling salesman problem show that the expected cost is very well predicted by this approach, but predictions of the correct minimum value of the cost function are very poor.

CHAPTER 3

Asymptotic Convergence

In the previous chapter we introduced the simulated annealing algorithm on the basis of the analogy with the simulation of the physical annealing process. We conjectured the existence of a stationary distribution for the simulated annealing algorithm equivalent to the existence of the Boltzmann distribution at thermal equilibrium in statistical physics. As a corollary of this conjecture we have demonstrated the asymptotic convergence of the simulated annealing algorithm to the set of globally optimal solutions; see Corollary 2.1.

In this chapter we prove the correctness of this conjecture. We introduce a mathematical formalism of the simulated annealing algorithm based on the theory of finite Markov chains by Feller [1950] and Seneta [1981], and discuss the conditions under which asymptotic convergence is guaranteed. We conclude this chapter with some comments on the asymptotic behaviour of the algorithm.

3.1 Markov Theory

As mentioned above, the simulated annealing algorithm can be modelled mathematically by using the theory of Markov chains.

Definition 3.1 A *Markov chain* is a sequence of *trials*, where the probability of the outcome of a given trial depends only on the outcome of the previous trial. Let $\mathbf{X}(k)$ be a stochastic variable denoting the outcome of the kth trial, then the *transition probability* at the kth trial for each pair i, j of outcomes is defined as

$$P_{ij}(k) = \mathbb{P}\{\mathbf{X}(k) = j | \mathbf{X}(k-1) = i\}. \tag{3.1}$$

The matrix $P(k)$ whose elements are given by (3.1) is called the *transition matrix*. ■

Let $a_i(k)$ denote the probability of outcome i at the kth trial, i.e.

$$a_i(k) = \mathbb{P}\{X(k) = i\}. \tag{3.2}$$

Then $a_i(k)$ is given by the following recursion:

$$a_i(k) = \sum_l a_l(k-1)P_{li}(k). \tag{3.3}$$

Definition 3.2 A Markov chain is called *finite* if it is defined on a finite set of outcomes. ∎

Definition 3.3 A Markov chain is called *inhomogeneous* if the associated transition probabilities depend on the trial number k. If the transition probabilities are independent of the trial number, the Markov chain is called *homogeneous*. ∎

Definition 3.4 A vector **a** is called *stochastic* if its components a_i satisfy the following conditions:

$$a_i \geq 0, \quad \text{for all } i, \quad \text{and} \quad \sum_i a_i = 1. \tag{3.4}$$

A matrix P is called *stochastic* if its components P_{ij} satisfy the following conditions:

$$P_{ij} \geq 0, \quad \text{for all } i, j, \quad \text{and} \quad \sum_j P_{ij} = 1, \quad \text{for all } i. \tag{3.5}$$

∎

In the simulated annealing algorithm, a trial corresponds to a transition, and the set of outcomes is given by the finite set of solutions. Clearly, in the case of the simulated annealing algorithm, the outcome of a trial depends only on the outcome of the previous trial; see Section 2.2. Consequently, we may use the concept of finite Markov chains.

Definition 3.5 (Transition probability) Let (S, f) be an instance of a combinatorial optimization problem. Then the *transition probabilities* for the simulated annealing algorithm are defined as follows:

$$\forall i, j \in S: \quad P_{ij}(k) = P_{ij}(c_k) = \begin{cases} G_{ij}(c_k)A_{ij}(c_k) & \text{if } i \neq j \\ 1 - \sum_{l \in S, l \neq i} P_{il}(c_k) & \text{if } i = j, \end{cases} \tag{3.6}$$

where $G_{ij}(c_k)$ denotes the *generation probability*, i.e. the probability of generating a solution j from a solution i, and $A_{ij}(c_k)$ denotes the *acceptance probability*, i.e. the probability of accepting the solution j, once it is generated from solution i. ∎

The $G_{ij}(c_k)$ and $A_{ij}(c_k)$ are conditional probabilities. The corresponding matrices $G(c_k)$ and $A(c_k)$ are called the *generation matrix* and *acceptance matrix*, respectively.

Hereafter, we use a special set of conditional probabilities given by the following definitions.

Definition 3.6 (Generation probability)

$$\forall i, j \in S: \quad G_{ij}(c_k) = G_{ij} = \frac{1}{\Theta} \chi_{(S_i)}(j), \tag{3.7}$$

where $\Theta = |S_i|$, for all $i \in S$. ∎

Definition 3.7 (Acceptance probability)

$$\forall i, j \in S: \quad A_{ij}(c_k) = \exp\left(-\frac{(f(j) - f(i))^+}{c_k}\right), \tag{3.8}$$

where, for all $a \in \mathbb{R}$, $a^+ = a$ if $a > 0$, and $a^+ = 0$ otherwise. ∎

Thus, the generation probabilities are chosen independent of the control parameter c_k and uniformly over the neighbourhoods S_i, where it is assumed that all neighbourhoods are of equal size, i.e. $|S_i| = \Theta$, for all $i \in S$. The acceptance probabilities are given by the acceptance criterion of (2.4) and is thus identical to the Metropolis criterion; see Section 2.1. Note furthermore that the transition and generation matrices for the simulated annealing algorithm are stochastic. The acceptance matrix however is not stochastic.

The definitions of the generation and acceptance probabilities, given by (3.7) and (3.8), correspond to the original definition of the simulated annealing algorithm and closely follow the physical analogy discussed in Chapter 2. These definitions can be used for virtually all combinatorial optimization problems, and close examination of the literature reveals that in most practical applications these definitions - or minor variations, which we will discuss at the end of Section 3.2 - are used indeed.

It is also possible to formulate a set of conditions ensuring asymptotic convergence for a more general class of acceptance and generation probabilities as was shown by Aarts & Van Laarhoven [1985a], Anily & Federgruen [1987b], Faigle & Schrader [1988], Van Laarhoven [1988], Lundy & Mees [1986], and Romeo & Sangiovanni-Vincentelli [1985]; see also Theorem 3.4 and Theorem 3.8. However, the probabilities of (3.7) and (3.8) are imposed by this class in a natural way [Aarts & Van Laarhoven, 1985a], and we therefore restrict ourselves here to these special probabilities.

We now concentrate on the asymptotic convergence properties of the simulated annealing algorithm. The simulated annealing algorithm finds with probability one an optimal solution if, after a possibly large number of trials, we have

$$\mathbb{P}\{\mathbf{X}(k) \in S_{opt}\} = 1. \tag{3.9}$$

In the following sections we show that under certain conditions the simulated annealing algorithm converges asymptotically to the set of optimal solutions, i.e.

$$\lim_{k \to \infty} \mathbb{P}\{\mathbf{X}(k) \in S_{opt}\} = 1. \tag{3.10}$$

3.2   The Stationary Distribution

Essential to the asymptotic convergence proof of the simulated annealing algorithm is the existence of a unique stationary distribution. Such a stationary distribution exists only under certain conditions on the Markov chains that can be associated with the algorithm. In this section we discuss these conditions and prove that, for the homogeneous Markov chains associated with the simulated annealing algorithm, the stationary distribution assumes the exponential form of (2.5), i.e. we prove Conjecture 2.1. For this we assume that the value of the control parameter is independent of k, i.e. $c_k = c$, for all k. We then have $P(k) = P$, for all k, which corresponds to a homogeneous Markov chain.

Definition 3.8 [Feller, 1950] The stationary distribution of a finite homogeneous Markov chain with transition matrix P is defined as the vector $\mathbf{q}$, whose i^{th} component is given by

$$q_i = \lim_{k \to \infty} \mathbb{P}\{\mathbf{X}(k) = i \mid \mathbf{X}(0) = j\}, \quad \text{for all } j. \tag{3.11}$$

∎

If such a stationary distribution $\mathbf{q}$ exists we have

$$\lim_{k \to \infty} a_i(k) = \lim_{k \to \infty} \mathbb{P}\{\mathbf{X}(k) = i\}$$

$$= \lim_{k \to \infty} \sum_j \mathbb{P}\{\mathbf{X}(k) = i | \mathbf{X}(0) = j\}\mathbb{P}\{\mathbf{X}(0) = j\}$$

$$= q_i \sum_j \mathbb{P}\{\mathbf{X}(0) = j\} = q_i. \tag{3.12}$$

Thus, the stationary distribution is the probability distribution of the solutions after an infinite number of trials. Furthermore, we have that

$$
\mathbf{q}^T = \lim_{k \to \infty} \mathbf{a}^T(0) \prod_{l=1}^{k} P(l) = \lim_{k \to \infty} \mathbf{a}^T(0) P^k
$$

$$
= \lim_{k \to \infty} \mathbf{a}^T(0) P^{k-1} P = \lim_{l \to \infty} \mathbf{a}^T(0) P^l P
$$

$$
= \mathbf{q}^T P. \tag{3.13}
$$

Thus, $\mathbf{q}$ is a left eigenvector of P with eigenvalue 1. Clearly, in the case of the simulated annealing algorithm, as P depends on c, $\mathbf{q}$ depends on c, i.e. $\mathbf{q} = \mathbf{q}(c)$.

Before we can prove the existance of the stationairy distribution for the simulated annealing algorithm, we need the following definitions.

Definition 3.9 A Markov chain with transition matrix P is *irreducible*, if for each pair of solutions $i, j \in S$ there is a positive probability of reaching j from i in a finite number of trials, i.e.

$$
\forall i, j \; \exists n \geq 1 : (P^n)_{ij} > 0. \tag{3.14}
$$

∎

Definition 3.10 A Markov chain with transition matrix P is *aperiodic*, if for each solution $i \in S$ the greatest common divisor $\gcd(D_i) = 1$, where the set D_i consists of all integers $n > 0$, with

$$
(P^n)_{ii} > 0. \tag{3.15}
$$

∎

The integer $\gcd(D_i)$ is called the *period* of solution i. Thus, aperiodicity requires all solutions to have period one.

Lemma 3.1 *An irreducible Markov chain with transition matrix P is aperiodic if*

$$
\exists j \in S : P_{jj} > 0. \tag{3.16}
$$

Proof Irreducibility implies that

$$
\forall i, j \in S, \; \exists k, l \geq 1 \; : \; (P^k)_{ij} > 0 \text{ and } (P^l)_{ji} > 0.
$$

Hence, $(P^n)_{ii} \geq (P^k)_{ij}(P^l)_{ji} > 0$, where $n = k + l$. Moreover, we have $(P^{n+1})_{ii} \geq (P^k)_{ij} P_{jj} (P^l)_{ji} > 0$. Thus, we have $n, n + 1 \in D_i$ and consequently $\gcd(D_i) = 1$, since $1 \leq \gcd(D_i) \leq \gcd(n, n+1)$, and $\gcd(n, n+1) = 1$, for all $n \geq 1$.   ■

Theorem 3.1 [Feller, 1950] *Let P be the transition matrix associated with a finite homogeneous Markov chain and let the Markov chain be irreducible and aperiodic. Then there exists a stochastic vector $\mathbf{q}$ whose components q_i are uniquely determined by the following equation:*

$$\sum_j q_j P_{ji} = q_i, \quad \text{for all } i. \tag{3.17}$$

■

Clearly, the vector $\mathbf{q}$ is the stationary distribution of the Markov chain, because it satisfies (3.13).

Lemma 3.2 *Let P be the transition matrix associated with a finite, irreducible and aperiodic homogeneous Markov chain. Then a given distribution is the stationary distribution of the Markov chain if its components satisfy the following equation:*

$$q_i P_{ij} = q_j P_{ji}, \quad \text{for all } i, j \in S. \tag{3.18}$$

Proof As a result of Theorem 3.1, the existence of a unique stationary distribution $\mathbf{q}$ is guaranteed. The proof of the theorem then is completed by proving that (3.18) implies (3.17), which is straightforward: for all $i, j \in S$ we have

$$q_i P_{ij} = q_j P_{ji}$$

$$\Rightarrow \quad \sum_{j \in S} q_i P_{ij} = \sum_{j \in S} q_j P_{ji}$$

$$\Rightarrow \quad q_i = \sum_{j \in S} q_j P_{ji},$$

since P is stochastic.   ■

Thus, the correctness of a given stationary distribution of a finite homogeneous Markov chain with transition matrix P can be easily verified by showing that the components of $\mathbf{q}$ satisfy (3.18), provided the Markov chain is irreducible and aperiodic. The expression (3.18) is called the *detailed balance equation*; similarly, the expression of (3.17) is called the *global balance equation*.

Theorem 3.2 *Let (S, f) denote an instance of a combinatorial optimization problem and $P(c)$ denote the transition matrix associated with the simulated annealing algorithm defined by (3.6), (3.7), and (3.8). Furthermore, let the* *following condition be satisfied:*

(1) $\forall i, j \in S \; \exists p \geq 1, \; \exists l_0, l_1, \ldots, l_p \in S,$

with $l_0 = i, l_p = j,$ and

$$G_{l_k l_{k+1}} > 0, \quad k = 0, 1, \ldots, p - 1. \tag{3.19}$$

Then the Markov chain has a stationary distribution $\mathbf{q}(c)$, *whose components are given by*

$$q_i(c) = \frac{1}{N_0(c)} \exp\left(-\frac{f(i)}{c}\right), \quad \text{for all } i \in S, \tag{3.20}$$

where

$$N_0(c) = \sum_{j \in S} \exp\left(-\frac{f(j)}{c}\right). \tag{3.21}$$

Proof To prove this theorem we first show that the Markov chain defined in this way is irreducible and aperiodic. Next, we apply Lemma 3.2 and verify the correctness of the stationary distribution by showing that the components given by (3.20) and (3.21) satisfy the detailed balance equation of (3.18).

Irreducibility: Using condition (1) we have

$$
\begin{aligned}
(P^p)_{ij}(c) &= \sum_{k_1 \in S} \sum_{k_2 \in S} \cdots \sum_{k_{p-1} \in S} P_{ik_1}(c) P_{k_1 k_2}(c) \cdots P_{k_{p-1} j}(c) \\
&\geq G_{il_1} A_{il_1}(c) G_{l_1 l_2} A_{l_1 l_2}(c) \cdots G_{l_{p-1} j} A_{l_{p-1} j}(c) \\
&> 0,
\end{aligned}
$$

since $A_{ij} > 0$, for all $i, j \in S$; cf. (3.8).

Aperiodicity: Let $i, j \in S$ with $f(i) < f(j)$ and $G_{ij} > 0$. Because of condition (1) such a pair i, j always exists provided $S \neq S_{opt}$. Then $A_{ij}(c) < 1$, and we thus have

$$
\begin{aligned}
P_{ii}(c) &= 1 - \sum_{l \in S, l \neq i} G_{il} A_{il}(c) \\
&= 1 - G_{ij} A_{ij}(c) - \sum_{l \in S, l \neq i, j} G_{il} A_{il}(c)
\end{aligned}
$$

$$> \quad 1 - G_{ij} - \sum_{l \in S, l \neq i, j} G_{il}$$

$$= \quad 1 - \sum_{l \in S, l \neq i} G_{il} = 0.$$

Thus, according to Theorem 3.1 there exists a unique stationary distribution, since the associated Markov chain is irreducible and aperiodic. Next, we apply Lemma 3.2 to complete the proof by showing that the stationary distribution defined by (3.20) and (3.21) is a vector with the following properties: (i) it is stochastic and (ii) its components satisfy the detailed balance equation given by (3.18).

(i) Trivial.

(ii) Note that the definition of the generation probabilities of (3.7) implies that $G_{ij} = G_{ji}$, since $\chi_{(S_j)}(i) = 1 \Leftrightarrow \chi_{(S_i)}(j) = 1$; see Definition 1.3. Hence, the detailed balance equation of (3.18) reduces to

$$q_i(c)A_{ij}(c) = q_j(c)A_{ji}(c), \quad \text{for all } i, j \in S. \tag{3.22}$$

Using Definition 3.7 and (3.20) and (3.21) we obtain

$$q_i(c)A_{ij}(c) = \frac{1}{N_0(c)} \exp\left(\frac{-f(i)}{c}\right) \exp\left(-\frac{(f(j) - f(i))^+}{c}\right)$$

$$= \frac{1}{N_0(c)} \exp\left(\frac{-f(j)}{c}\right) \exp\left(-\frac{(f(i) - f(j)) + (f(j) - f(i))^+}{c}\right)$$

$$= \frac{1}{N_0(c)} \exp\left(\frac{-f(j)}{c}\right) \exp\left(-\frac{(f(i) - f(j))^+}{c}\right)$$

$$= q_j(c)A_{ji}(c),$$

which completes the proof of the theorem. ∎

The stationary distribution defined by (3.20) and (3.21) is identical to the distribution of (2.5) and (2.6), which was conjectured from the physical analogy. From Corollary 2.1 we recall that

$$\lim_{c \downarrow 0} \mathbf{q}(c) = \mathbf{q}^*, \tag{3.23}$$

where the components of $\mathbf{q}^*$ are given by

$$q_i^* = \frac{1}{|S_{opt}|} \chi_{(S_{opt})}(i). \tag{3.24}$$

Hence, we finally obtain

$$\lim_{c \downarrow 0} \lim_{k \to \infty} \mathbf{P}_c\{\mathbf{X}(k) = i\} = \lim_{c \downarrow 0} q_i(c) = q_i^*, \tag{3.25}$$

or

$$\lim_{c \downarrow 0} \lim_{k \to \infty} \mathbf{P}_c\{\mathbf{X}(k) \in S_{opt}\} = 1. \tag{3.26}$$

This result reflects the basic property of the simulated annealing algorithm, i.e. the guarantee that the algorithm asymptotically finds an optimal solution.

We end this section with some remarks. The convergence proof given above remains essentially unchanged upon replacing Definition 3.6 by the more general requirement that the generation probabilities be symmetric, i.e.

$$G_{ij} = G_{ji}, \quad \text{for all } i, j \in S. \tag{3.27}$$

Similarly, Lundy & Mees [1986] show that the generation probabilities can be defined as

$$G_{ij} = \frac{1}{|S_i|} \chi_{(S_i)}(j), \quad \text{for all } i, j \in S. \tag{3.28}$$

In this case the components of the stationary distribution are given by

$$q_i(c) = \frac{|S_i| \exp\left(\frac{-f(i)}{c}\right)}{\sum_{j \in S} |S_j| \exp\left(\frac{-f(j)}{c}\right)}. \tag{3.29}$$

These components still converge to the q_i^* given by (3.24) as $c \downarrow 0$.

Finally, we mention that a generation matrix satisfying both (3.27) and (3.28) implies that $|S_i| = \Theta$, for all $i \in S$. This can be straightforwardly shown.

Example 3.1 (Travelling salesman problem; see Examples 1.1 and 1.3) Let (S, f) denote an instance of a travelling salesman problem with n cities and let $\mathcal{N}_2$ denote the 2-change neighbourhood structure. Then according to (3.7) and (1.9) we have

$$G_{ij} = \frac{1}{(n-1)(n-2)} \chi_{(S_i)}(j). \tag{3.30}$$

Furthermore, as a result of the 2-change mechanism, condition (1) of Theorem 3.2 is satisfied; cf. Example 1.3. Consequently, the simulated annealing algorithm will asymptotically find an optimal tour, when applied to the travelling salesman problem using the 2-change neighbourhood structure with the generation probabilities of (3.30) and the acceptance probabilities of (3.8). ∎

As mentioned in Section 3.1, asymptotic convergence of the simulated annealing algorithm can also be proved for a more general class of generation and acceptance probabilities. This is stated in the following theorem.

Theorem 3.3 *Let* (S, f) *denote an instance of a combinatorial optimization problem, let the transition probabilities associated with the simulated annealing algorithm be defined by (3.6) and let the generation probabilities* $G_{ij}(c)$ *and acceptance probabilities* $A_{ij}(c)$ *satisfy the following conditions:*

$(G1)$ $\quad \forall c > 0 \ \forall i, j \in S \ \exists p \geq 1, \ \exists l_0, l_1, \ldots, l_p \in S, with \ l_0 = i, l_p = j,$
$$and \ G_{l_k l_{k+1}}(c) > 0, \quad k = 0, 1, \ldots, p - 1,$$

$(G2)$ $\qquad\qquad \forall c > 0, \ \forall i, j \in S : \ G_{ij}(c) = G_{ji}(c),$

$(A1)$ $\quad \forall c > 0, \ \forall i, j \in S : \quad A_{ij}(c) = 1 \qquad if \ f(i) \geq f(j),$
$$A_{ij}(c) \in (0, 1) \quad if \ f(i) < f(j),$$

$(A2)$ $\qquad \forall c > 0, \ \forall i, j, k \in S \ with \ f(i) \leq f(j) \leq f(k) :$
$$A_{ik}(c) = A_{ij}(c) A_{jk}(c),$$

$(A3)$ $\qquad \forall i, j \in S \ with \ f(i) < f(j) : \ \lim_{c \downarrow 0} A_{ij}(c) = 0.$

Then there exists a stationary distribution $\mathbf{q}(c)$ *whose components are given by*

$$q_i(c) = \frac{A_{i_{opt} i}(c)}{\sum_{j \in S} A_{i_{opt} j}(c)}, \quad for \ all \ i \in S, \tag{3.31}$$

and for arbitrary $i_{opt} \in S$. *Moreover we have*

$$\lim_{c \downarrow 0} q_i(c) = \frac{1}{|S_{opt}|} \chi_{(S_{opt})}(i). \tag{3.32}$$

Proof The proof of this theorem is similar to the proofs of Theorem 3.2 and Corollary 2.1; further details are given in [Van Laarhoven & Aarts, 1987]. ∎

The set of conditions given in Theorem 3.3 is, except for condition (G1), the most general set of conditions on the generation and acceptance probabilities known in the literature. Although the conditions (A1)-(A3) of Theorem 3.3 define a more general class of acceptance probabilities than the exponential form given in (3.8), it is very hard to construct analytical expressions for the acceptance probabilities that satisfy (A1)-(A3) but which are essentially different from the exponential form of (3.8). To the best of our knowledge, the literature does not give any examples of alternative expressions.

As mentioned above, condition (G1) of Theorem 3.3 can be replaced by a more general condition. It should be noted that this condition, similar to condition (1) of Theorem 3.2, states that the Markov chain associated with the generation matrix G is irreducible in itself. If this is not the case, asymptotic convergence to a subset of the set of globally optimal solutions can still be proved [Feller, 1950] if condition (G1) in Theorem 3.3 is replaced by the following condition, which is both necessary and sufficient:

$$(G1') \qquad \forall i \in S \; \exists i_{opt}, p \geq 1, \; \exists l_0, l_1, \ldots, l_p \in S,$$

$$\text{with } l_0 = i, l_p = i_{opt}, \text{ and}$$

$$G_{l_k l_{k+1}} > 0, \quad k = 0, 1, \ldots, p - 1. \tag{3.33}$$

Condition (G1') states that it must be possible to construct a finite sequence of transitions with non-zero generation probability, leading from an arbitrary solution i to some optimal solution i_{opt}.

For the proof of the validity of condition (G1'), a distinction must be made between *transient* and *recurrent* solutions, where a solution is called transient if the probability that the Markov chain ever returns to that solution equals zero, and recurrent if the Markov chain may return to the solution with a finite probability [Feller, 1950]. Furthermore, we mention that in this case the stationary distribution of (3.31) does not exist anymore and should be replaced by a *stationary matrix* $Q(c)$ whose elements q_{ij} denote the probability of finding a solution j after an infinite number of transitions, starting off from a solution i. A more detailed treatment of this is considered to be beyond the scope of this book. For a detailed treatment of this subject the reader is referred to [Connors & Kumar, 1987; Gidas, 1985a; Van Laarhoven, 1988; Van Laarhoven, Aarts & Lenstra, 1988].

3.3 Inhomogeneous Markov Chains

The asymptotic convergence proofs, given in the previous section, indicate that the simulated annealing algorithm requires an infinite number of transitions to approximate a stationary distribution arbitrarily close. A more precise bound on the number of transitions is given in Section 3.4. Thus, implementation of the algorithm along these lines would require generation of a sequence of infinitely long homogeneous Markov chains at descending values of the control parameter c; cf. (3.26). This is clearly impracticable. However, we can do slightly better, i.e. we can describe the simulated annealing algorithm as a sequence of homogeneous Markov chains of finite length, generated at descending values of the control parameter. This process can be described by combining the homogeneous Markov chains into one single inhomogeneous Markov chain. In this way the sequence of infinitely long homogeneous Markov chains is reduced to one single inhomogeneous Markov chain of infinite length.

Definition 3.11 Let c_l' denote the value of the control parameter of the lth homogeneous Markov chain, L denote the length of the homogeneous Markov chains, and c_k denote the value of the control parameter at the kth trial. Then we define the sequence $\{c_k\}$, $k = 1, \ldots$ as follows:

$$c_k = c_l', \quad lL < k \leq (l+1)L. \tag{3.34}$$

Thus, the control parameter is taken to be piecewise constant. Furthermore, the sequence $\{c_l'\}$, $l = 0, 1, \ldots$ is defined such that it satisfies the following conditions:

$$c_{l+1}' \leq c_l', \quad l = 0, 1, \ldots \tag{3.35}$$

and

$$\lim_{l \to \infty} c_l' = 0. \tag{3.36}$$

∎

Clearly, the lengths of the homogeneous Markov chains could have been chosen variable, which would have been more general. However, a fixed length imposes no major restrictions; it merely simplifies the calculations.

In Section 3.4 it will be proved that, under certain conditions on the rate of convergence of the sequence $\{c_l'\}$, the inhomogeneous Markov chain associated with the simulated annealing algorithm converges in distribution to the distribution $\mathbf{q}^*$, whose components are given by (3.24). In other words we prove that

$$\lim_{k \to \infty} \mathbb{P}\{\mathbf{X}(k) = i\} = q_i^* = \frac{1}{|S_{opt}|} \mathcal{X}(S_{opt})(i). \tag{3.37}$$

Before we can prove this we need the following definitions.

Definition 3.12 Let $P(k)$ be the transition matrix associated with an inhomogeneous Markov chain. Then the matrix $U(m, n)$ is defined as

$$U(m, n) = \prod_{k=m}^{n} P(k), \quad 0 < m \leq n. \tag{3.38}$$

In other words the the components of $U(m, n)$ are equal to

$$U_{ij}(m, n) = \mathbb{P}\{X(n) = j | X(m - 1) = i\}. \tag{3.39}$$

∎

Definition 3.13 [Seneta, 1981] Let P be an $n \times n$ stochastic matrix, then the *coefficient of ergodicity* $\tau_1(P)$ is defined as

$$\begin{aligned}
\tau_1(P) &= \frac{1}{2} \max_{i,j} \sum_{k=1}^{n} |P_{ik} - P_{jk}| \\
&= 1 - \min_{i,j} \sum_{k=1}^{n} \min(P_{ik}, P_{jk}).
\end{aligned} \tag{3.40}$$

∎

Definition 3.14 [Seneta, 1981] A finite inhomogeneous Markov chain is *weakly ergodic* if

$$\forall i, j, l \in S, \forall m > 0: \lim_{k \to \infty} (U_{il}(m, k) - U_{jl}(m, k)) = 0. \tag{3.41}$$

∎

Definition 3.15 [Seneta, 1981] A finite inhomogeneous Markov chain is *strongly ergodic* if there exists a stochastic vector $\mathbf{q}^*$, such that

$$\forall i, j \in S, \forall m > 0: \lim_{k \to \infty} U_{ij}(m, k) = q_j^*. \tag{3.42}$$

∎

Thus, for a given m, weak ergodicity implies that $X(k)$ becomes independent of $X(m)$, as $k \to \infty$, whereas strong ergodicity implies *convergence in distribution*, i.e.

$$\lim_{k \to \infty} \mathbf{a}^T(m - 1) \prod_{n=m}^{k} P(n) = (\mathbf{q}^*)^T, \tag{3.43}$$

or

$$\lim_{k \to \infty} \mathbb{P}\{X(k) = j\}$$

$$= \lim_{k \to \infty} (\sum_{i \in S} \mathbb{P}\{\mathbf{X}(k) = j | \mathbf{X}(m-1) = i\} \mathbb{P}\{\mathbf{X}(m-1) = i\})$$

$$= \lim_{k \to \infty} (\sum_{i \in S} U_{ij}(m, k) \mathbb{P}\{\mathbf{X}(m-1) = i\})$$

$$= q_j^* (\sum_{i \in S} \mathbb{P}\{\mathbf{X}(m-1) = i\}) = q_j^*. \tag{3.44}$$

Note that, for a homogeneous Markov chain, there is no distinction between weak and strong ergodicity.

3.4 Convergence in Distribution

The following two theorems provide conditions for weak and strong ergodicity of inhomogeneous Markov chains.

Theorem 3.4 [Seneta, 1981] *An inhomogeneous Markov chain is weakly ergodic if and only if there is a strictly increasing sequence of positive numbers* $\{k_i\}$, $i = 0, 1, 2, \ldots$, *such that*

$$\sum_{i=0}^{\infty} (1 - \tau_1(U(k_i, k_{i+1}))) = \infty. \tag{3.45}$$ ∎

Theorem 3.5 [Isaacson & Madsen, 1976] *A finite inhomogeneous Markov chain is strongly ergodic under the following conditions:*

(*i*) *the Markov chain is weakly ergodic,*

(*ii*) *for all k there exists a stochastic vector* $\mathbf{q}(k)$ *such that* $\mathbf{q}(k)$ *is the left eigenvector of* $P(k)$ *with eigenvalue* 1.

(*iii*) *the eigenvectors* $\mathbf{q}(k)$ *satisfy*

$$\sum_{k=1}^{\infty} \|\mathbf{q}(k) - \mathbf{q}(k+1)\| < \infty. \tag{3.46}$$

Moreover, if $\mathbf{q}^* = \lim_{k \to \infty} \mathbf{q}(k)$, *then* $\mathbf{q}^*$ *is the vector in Definition 3.15.* ∎

Convergence in distribution can now be proved by showing that the inhomogeneous Markov chain associated with the simulated annealing algorithm is strongly ergodic. This is the subject of the next theorem.

Theorem 3.6 *Let* (S, f) *denote an instance of a combinatorial optimization*

problem and let $P(k)$ denote the transition matrix associated with the simu-
lated annealing algorithm defined by (3.6), (3.7), and (3.8). Furthermore, let
the following conditions be satisfied:

(1)
$$\forall i, j \in S \; \exists p \geq 1, \; \exists l_0, l_1, \ldots, l_p \in S,$$

with $l_0 = i, l_p = j$, and

$$G_{l_k l_{k+1}} > 0, \quad k = 0, 1, \ldots, p - 1. \tag{3.47}$$

(2) The sequence $\{c_l'\}$ satisfies the following inequality:

$$c_l' \geq \frac{(L+1)\Delta}{\log(l+2)}, \quad l = 0, 1, \ldots, \tag{3.48}$$

where

$$\Delta = \max_{i,j \in S}\{f(j) - f(i) | j \in S_i\} \tag{3.49}$$

and L is chosen as the maximum of the minimum number of transitions
required to reach an i_{opt} from j, for all $j \in S$; such an L always exists
as a result of condition (1).

Then the Markov chain converges in distribution to the vector $\mathbf{q}^*$, with com-
ponents

$$q_i^* = \frac{1}{|S_{opt}|} \chi_{(S_{opt})}(i), \quad \text{for all } i \in S, \tag{3.50}$$

or, in other words

$$\lim_{k \to \infty} \mathbb{P}\{\mathbf{X}(k) \in S_{opt}\} = 1. \tag{3.51}$$

Proof To prove this theorem we use Theorem 3.5 and show that conditions (1)
and (2) are sufficient to satisfy conditions (i), (ii) and (iii) of Theorem 3.5.

(i) The following inequalities can be straightforwardly proved:

(A) $\quad \forall k > 0, \forall i \in S, j \in S_i \backslash \{i\}: \quad P_{ij}(k) \geq \dfrac{1}{\Theta} \exp\left(\dfrac{-\Delta}{c_k}\right)$

and

(B) $\quad \exists k_0 > 0 : \forall k > k_0, \forall i \in S \backslash \hat{S}': \quad P_{ii}(k) \geq \dfrac{1}{\Theta} \exp\left(\dfrac{-\Delta}{c_k}\right),$

where Δ is given by (3.49) and $\hat{S}'$ denotes the set of locally maximal
solutions. Indeed, (A) follows directly from the definition of the $P_{ij}(k)$

and Δ, given in (3.6), (3.7), and (3.8) and (3.49), respectively. (B) is obtained as follows. Let $\delta = \min_{i,j\in S}^{+}\{f(j) - f(i)|j \in S_i\}$, where $\min^{+}$ denotes that the minimum is taken over the positive terms only. Then it is always possible to find a $k_0 > 0$ such that

$$1 - \exp\left(\frac{-\delta}{c_k}\right) > \exp\left(\frac{-\Delta}{c_k}\right), \quad \text{for all } k > k_0,$$

Consequently, it follows directly that

$$\forall k > k_0, \ \forall i \in S\backslash\hat{S}' \ : \ P_{ii}(k) = 1 - \sum_{j\in S_i\backslash\{i\}} P_{ij}(k)$$

$$= 1 - \sum_{j\in S_i\backslash\{i\}} \frac{1}{\Theta}A_{ij}(c_k) \geq 1 - \frac{1}{\Theta}\left\{\Theta - 1 - \exp\left(\frac{-\delta}{c_k}\right)\right\}$$

$$= \frac{1}{\Theta}\left\{1 - \exp\left(\frac{-\delta}{c_k}\right)\right\} > \frac{1}{\Theta}\exp\left(\frac{-\Delta}{c_k}\right).$$

Now let $p \in S$ be such that $U_{pi_{opt}}(k, k+L) = \min_{i\in S} U_{ii_{opt}}(k, k+L)$. Then, as a result of the choice of L, we have for all $k \geq k_0$

$$\min_{i\in S} U_{ii_{opt}}(k, k+L)$$

$$= \mathbb{P}\{\mathbf{X}(k+L) = i_{opt} \mid \mathbf{X}(k-1) = p\}$$

$$= \sum_{j_1\in S}\sum_{j_2\in S}\cdots\sum_{j_L\in S} P_{pj_1}(k)P_{j_1 j_2}(k+1)\cdots P_{j_L i_{opt}}(k+L)$$

$$\geq P_{pl_1}(k)P_{l_1 l_2}(k+1)\cdots P_{l_L i_{opt}}(k+L)$$

$$\geq \left[\frac{1}{\Theta}\exp\left(\frac{-\Delta}{c_k}\right)\right]^{L+1}. \tag{3.52}$$

Hence, we obtain:

$$1 - \tau_1(U(k, k+L))$$

$$= \min_{i,j\in S}\sum_{p\in S} \min\{U_{ip}(k, k+L), U_{jp}(k, k+L)\}$$

$$\geq \min_{i\in S} U_{ii_{opt}}(k, k+L)$$

$$\geq \Theta^{-L-1} \exp\left(\frac{-(L+1)\Delta}{c_k}\right), \quad \text{for all } k \geq k_0.$$

Weak ergodicity now follows directly from Theorem 3.4 by setting $k_l = lL$, such that we obtain

$$\sum_{l=0}^{\infty}(1 - \tau_1(U(k_l, k_{l+1}))) \geq \sum_{l=k_0}^{\infty}(1 - \tau_1(U(lL, lL+L)))$$

$$\geq \sum_{l=k_0}^{\infty} \Theta^{-L-1} \exp\left(\frac{-(L+1)\Delta}{c_l'}\right)$$

$$\geq \sum_{l=k_0}^{\infty} \Theta^{-L-1}(l+2)^{-1} = \infty.$$

(ii) Condition (1) guarantees the existence of the left eigenvector $\mathbf{q}(k)$ of $P(k)$, given by $\mathbf{q}(k) = \mathbf{q}(c_k)$, i.e. the stationary distribution of the homogeneous Markov chain with transition matrix $P = P(k)$; see Theorem 3.2. For the transition matrix associated with the simulated annealing algorithm defined by (3.6), (3.7) and (3.8), the components of the eigenvectors are given by (3.20) and (3.21).

(iii) From Corollary 2.4 we have

$$\forall i \in S_{opt} \quad : \quad q_i(c_{k+1}) - q_i(c_k) > 0$$

and

$$\exists k_1 : \forall k > k_1, \ \forall i \in S \backslash S_{opt} \quad : \quad q_i(c_{k+1}) - q_i(c_k) < 0.$$

Consequently, we obtain

$$\forall k \geq k_1 \quad : \quad ||\mathbf{q}(k) - \mathbf{q}(k+1)|| \overset{\text{def}}{=} \sum_{i \in S}|q_i(c_k) - q_i(c_{k+1})|$$

$$= \sum_{i \in S_{opt}}(q_i(c_{k+1}) - q_i(c_k)) - \sum_{i \in S \backslash S_{opt}}(q_i(c_{k+1}) - q_i(c_k)).$$

Substituting $\sum\limits_{i \in S \backslash S_{opt}} q_i(c) = 1 - \sum\limits_{i \in S_{opt}} q_i(c)$ then yields

$$\forall k > k_1 \quad : \quad ||\mathbf{q}(k) - \mathbf{q}(k+1)|| = 2\left(\sum_{i \in S_{opt}} q_i(c_{k+1}) - \sum_{i \in S_{opt}} q_i(c_k)\right).$$

We then obtain

$$\sum_{k=k_1+1}^{\infty} ||\mathbf{q}(k) - \mathbf{q}(k+1)|| = 2 \left(\sum_{i \in S_{opt}} q_i(c_\infty) - \sum_{i \in S_{opt}} q_i(c_{k_1+1}) \right) \leq 2.$$

Thus, (3.46) can now be written as

$$\sum_{k=1}^{\infty} ||\mathbf{q}(k) - \mathbf{q}(k+1)||$$

$$= \sum_{k=1}^{k_1} ||\mathbf{q}(k) - \mathbf{q}(k+1)|| + \sum_{k=k_1+1}^{\infty} ||\mathbf{q}(k) - \mathbf{q}(k+1)|| \leq 2k_1 + 2.$$

Thus, the sum $\sum_{k=1}^{\infty} ||\mathbf{q}(k) - \mathbf{q}(k+1)||$ is finite, which completes the proof of part (iii).

Finally, we know from Corollary 2.1 that $\lim_{k\to\infty} \mathbf{q}(k) = \mathbf{q}^*$, where the components of the vector $\mathbf{q}^*$ are given by (3.50) since $\lim_{k\to\infty} c_k = \lim_{l\to\infty} c'_l = 0$. This completes the proof of the theorem.   ∎

In conclusion, we have shown that the simulated annealing algorithm, formulated as a sequence of homogeneous Markov chains of finite length, converges in distribution to the set of optimal solutions, provided the cooling is done sufficiently slowly; see condition (2) of Theorem 3.6.

Note that the conditions for asymptotic convergence given above are *sufficient* but not *necessary*. Similar conditions have been derived by a number of authors. We mention Anily & Federgruen [1987a; 1987b], Gelfand & Mitter [1985], Geman & Geman [1984], Gidas [1985a; 1985b], Holley & Stroock [1988], and Mitra, Romeo & Sangiovanni-Vincentelli [1986].

Necessary and sufficient conditions are derived by Hajek [1988]. The essential difference between the various results lies in the difference in value of the constants that play a role similar to that of the constant Δ in (3.48). The sharpest bound is given by Hajek [1988], which is evident since the conditions are both necessary and sufficient. To discuss this result we need some definitions.

Definition 3.16 Let $i, j \in S$, then j is *reachable at height h* from i if $\exists p \geq 1$, $\exists l_0, \ldots, l_p \in S$ with $l_0 = i$ and $l_p = j$, such that $G_{l_k, l_{k+1}} > 0$ and $f(l_k) \leq h$, for all $k = 0, \ldots, p - 1$.   ∎

Definition 3.17 Let $\hat{\imath}$ be a local minimum, then the *depth* $d(\hat{\imath})$ of $\hat{\imath}$ is the smallest h such that there is a solution $j \in S$ with $f(j) < f(\hat{\imath})$ that is reachable at height $f(\hat{\imath}) + h$ from $\hat{\imath}$. By definition $d(i_{opt}) = \infty$.   ■

We now discuss the results obtained by Hajek.

Theorem 3.7 [Hajek, 1988] *Let $\{c_k\}$ be a sequence of values of the control parameter similar to the sequence given in (3.48), i.e.*

$$c_k = \frac{\Gamma}{\log(k+2)}, \quad k, = 0, 1, \ldots, \qquad (3.53)$$

for some constant Γ. Then asymptotic convergence of the simulated annealing algorithm, using the transition probabilities of (3.6), (3.7) and (3.8), is guaranteed if and only if

   (*i*)   *condition (1) of Theorem 3.6 holds,*

  (*ii*)   *i is reachable from j at height h if and only if j is reachable from i at height h, for arbitrary $i, j \in S$ and h, and*

 (*iii*)   *the constant Γ satisfies $\Gamma \geq D$, where*

$$D = \max_{\hat{\imath} \in S \backslash S_{opt}} d(\hat{\imath}), \qquad (3.54)$$

i.e. D is the depth of the deepest local, non global minimum.   ■

From the fact that Hajek's conditions are necessary and sufficient we conclude that $D \leq \Delta$ where Δ is given by (3.49).

   Bounds on the value of D have been derived by Kern [1986b] for several combinatorial optimization problems. In particular, Kern showed, for a number of problems, that it is very unlikely that D can be calculated in polynomial time for each instance of a combinatorial optimization problem.

   Finally, we mention that, under certain conditions, asymptotic convergence of the inhomogeneous Markov chain associated with the simulated annealing algorithm can also be proved for general conditions on the generation and acceptance probabilities; see also Theorem 3.3. This result was first proved by Anily & Federgruen [1987b] and can be formulated as given below.

Theorem 3.8 [Anily & Federgruen, 1987b] *Let the transition probabilities associated with the simulated annealing algorithm be defined by (3.6) and let the generation probabilities $G_{ij}(c)$ and acceptance probabilities $A_{ij}(c)$ satisfy conditions (G1)-(A3) of Theorem 3.3. Furthermore, let*

$$\underline{A}(c) = \min_{i,j}\{A_{ij}(c) \mid i \in S, j \in S_i\}.$$

Then

$$\lim_{k \to \infty} \mathbb{P}\{\mathbf{X}(k) \in S_{opt}\} = 1,$$

if

$$\sum_{k=0}^{\infty} \left(\underline{A}(c_k')\right)^{L+1} = \infty, \tag{3.55}$$

where the sequence $\{c_k'\}$ is defined by (3.34) and L denotes the minimal number of transitions to reach an optimal solution from any given solution; see also condition (2) of Theorem 3.6. ∎

Applying this result to the special case where the acceptance probabilities are defined by (3.8) yields that asymptotic convergence to the set of globally optimal solutions is obtained if the sequence $\{c_k'\}$ satisfies condition (2) of Theorem 3.6.

3.5 Asymptotic Behaviour

In the previous sections we have shown that the simulated annealing algorithm converges in probability to the set of optimal solutions, or in other words, asymptotically the algorithm finds an optimal solution with probability one. As a result of the limits of (3.26) or (3.51), asymptotic convergence to the set of optimal solutions is achieved only after an infinite number of transitions. In any practical implementation this is clearly impracticable and one thus must resort to an approximation of the asymptotic convergence.

Below, we discuss some properties that can be derived for the *asymptotic behaviour* of the simulated annealing algorithm based on the mathematical framework presented in the previous sections. The first two properties that are discussed relate to the approximation of the stationary distribution. The third property relates to the approximation of the uniform distribution $\mathbf{q}^*$ on the set of globally optimal solutions.

With respect to the approximation of the stationary distribution we have the following two properties.

Property 3.1 [Seneta, 1981] Let $P(c)$ denote the transition matrix of the homogeneous Markov chain associated with the simulated annealing algorithm defined by (3.6) and $\mathbf{q}(c)$ denote the corresponding stationary distribution given by the left eigenvector with eigenvalue 1 of $P(c)$. Then, as $k \to \infty$, we have

$$\|\mathbf{a}(k) - \mathbf{q}(c)\| = \mathcal{O}(k^s |\lambda_2(c)|^k), \tag{3.56}$$

where $\lambda_2(c)$ $(0 < |\lambda_2(c)| < 1)$ denotes the second largest eigenvalue of $P(c)$ with multiplicity m_2, and $s = m_2 - 1$. ∎

Hence, the speed of convergence to the stationary distribution is determined by $\lambda_2(c)$. Unfortunately, computation of $\lambda_2(c)$ is impracticable, due to large size of the matrix $P(c)$. Approximation of the norm in (3.56) leads to the following property.

Property 3.2 [Aarts & Van Laarhoven, 1985a] Let ε denote an arbitrarily small positive number, then

$$\|\mathbf{a}(k) - \mathbf{q}(c)\| < \varepsilon \tag{3.57}$$

if

$$k > K \left(1 + \frac{\ln(\frac{1}{2}\varepsilon)}{\ln(1 - \gamma^K(c))}\right), \tag{3.58}$$

where $\gamma(c) = \min^+_{i,j\in S} P_{ij}(c)$ and $K = |S|^2 - 3|S| + 3$. ∎

Hence, (3.57) and (3.58) indicate that the stationary distribution is approximated arbitrarily closely, only if the number of transitions is at least quadratic in the size of the solution space. Moreover, this size $|S|$ is for most problems exponential in the size of the problem itself, for instance in the travelling salesman problem $|S| = (n - 1)!$, where n denotes the number of cities. Thus, the analysis presented above indicates that approximating the stationary distribution arbitrarily closely results in an exponential-time execution of the simulated annealing algorithm.

With respect to the asymptotic convergence of the inhomogeneous Markov chain associated with the simulated annealing algorithm we have the following result.

Property 3.3 [Mitra, Romeo & Sangiovanni-Vincentelli, 1986] Let the transition matrix of the inhomogeneous Markov chain associated with the simulated annealing algorithm be defined by (3.6), (3.7) and (3.8), and let the sequence $\{c'_l\}$ be given by (3.48), i.e.

$$c'_l = \frac{(L + 1)\Delta}{\log(l + 2)}, \quad l = 0, 1, \ldots, \tag{3.59}$$

where L and Δ are defined as in Theorem 3.6,[1] and let $\mathbf{q}^*$ be the uniform probability distribution on the set of optimal solutions defined by (3.50). Then

---

[1] The definitions of L and Δ, used by Mitra, Romeo & Sangiovanni-Vincentelli in the original version of Property 3.3, are slightly different from the ones used in Theorem 3.6. The proof of the property however is quite similar.

as $k \rightarrow \infty$, we have

$$\|\mathbf{a}(k) - \mathbf{q}^*\| < \varepsilon, \tag{3.60}$$

for an arbitrarily small positive number ε, if

$$k = \mathcal{O}\left(\left(\frac{1}{\varepsilon}\right)^{\frac{1}{\min(a,b)}}\right), \tag{3.61}$$

where

$$a = \frac{1}{(L+1)\Theta^{L+1}} \quad \text{and} \quad b = \frac{\hat{f} - f_{opt}}{(L+1)\Delta}, \tag{3.62}$$

with $\hat{f} = \min\limits_{i \in S \setminus S_{opt}} f(i)$.   ∎

Evaluation of this bound for particular problem instances typically leads to a number of transitions that is larger than the size of the solution space and thus to an exponential-time execution for most problems. This is illustrated by the following example.

Example 3.2 (Travelling salesman problem; see Examples 1.1 and 1.3) Let (S, f) denote an instance of a travelling salesman problem with n cities and $\mathcal{N}_2$ denote the 2-change neighbourhood structure. Then according to Example 1.3 we have $L = n - 2$ and $\Theta = (n-1)(n-2)$. Hence, (3.62) yields

$$a \approx \frac{1}{n-1}\left(\frac{1}{(n-1)(n-2)}\right)^{n-1} \quad \text{and} \quad b < \frac{1}{n-1}, \tag{3.63}$$

where n denotes the number of cities. Next, by using that $a \ll b$ and by choosing $\frac{1}{\varepsilon} = n$ we obtain from (3.61) that

$$k = \mathcal{O}\left(n^{n^{2n-1}}\right), \tag{3.64}$$

whereas $|S| = \mathcal{O}((n-1)!)$. Hence, complete enumeration of all solutions would take less time than approximating an optimal solution arbitrarily closely by the simulated annealing algorithm.   ∎

Summarizing, we have shown that the simulated annealing algorithm behaves as an optimization algorithm only if it is allowed an infinite number of transitions. Approximating the asymptotic behaviour arbitrarily closely, requires a number of transitions that for most problems is typically larger than the size of the solution space, leading to an exponential-time execution of the algorithm. Thus, the simulated annealing algorithm is clearly unsuited for solving combinatorial optimization problems to optimality. However, in the next chapter we will show how the asymptotic behaviour of the simulated annealing algorithm can be approximated in polynomial time. Evidently, this is at the cost of the guarantee of obtaining optimal solutions, but, as will be shown, the simulated annealing algorithm using this polynomial-time approximation returns near-optimal solutions for most problem instances.

CHAPTER 4

Finite-Time Approximation

In the previous chapter we have shown that implementation of the simulated annealing algorithm as an optimization algorithm requires an infinite number of transitions. In this chapter we focus on the finite-time implementation of the algorithm resulting in approximation of an optimal solution of the combinatorial optimization problem at hand.

4.1 Cooling Schedules

A *finite-time* implementation of the simulated annealing algorithm can be realized by generating homogeneous Markov chains of finite length for a finite sequence of descending values of the control parameter. To achieve this one must specify a set of parameters that governs the convergence of the algorithm. These parameters are combined in a so-called cooling schedule.

Definition 4.1 A *cooling schedule* specifies

- a finite sequence of values of the control parameter, i.e.

    - an *initial value* of the control parameter c_0,

    - a *decrement function* for decreasing the value of the control parameter,

    - a *final value* of the control parameter specified by a *stop criterion*, and

- a finite number of transitions at each value of the control parameter, i.e.

    – a finite *length* of each homogeneous Markov chain. ∎

In this section we discuss some general features and characteristics of cooling schedules. Furthermore, we present a conceptually simple cooling schedule that is frequently used in the literature.

Central in the discussion of cooling schedules is the concept of quasi equilibrium, which can be defined as follows.

Definition 4.2 Let L_k denote the length of the kth Markov chain and c_k the corresponding value of the control parameter. Then *quasi equilibrium* is achieved if $\mathbf{a}(L_k, c_k)$, i.e. the probability distribution of the solutions after L_k trials of the kth Markov chain, is 'sufficiently close' to $\mathbf{q}(c_k)$, the stationary distribution at c_k, defined by (3.20) and (3.21), i.e.

$$||\mathbf{a}(L_k, c_k) - \mathbf{q}(c_k)|| < \varepsilon, \qquad (4.1)$$

for some specified positive value of ε. ∎

We recall from the previous chapter that requiring (4.1) to hold for arbitrarily small values of ε implies that a number of transitions is needed which is quadratic in the size of the solution space. This leads for most combinatorial optimization problems to an exponential-time execution of the simulated annealing algorithm. Thus, for practical application of the algorithm, we need a less rigid quantification of quasi equilibrium than that of condition (4.1). In the literature this has led to different interpretations of the concept of quasi equilibrium, resulting in a rich variety of cooling schedules; cf. Van Laarhoven & Aarts [1987]. We give a more precise quantification in the next section.

The construction of a cooling schedule using the concept of quasi equilibrium is based on the following arguments. Let the acceptance probability be defined by (3.7), then, for $c \to \infty$, the stationary distribution is given by the uniform distribution on the set of solutions S, i.e.

$$\lim_{c \to \infty} \mathbf{q}(c) = \frac{1}{|S|} \mathbf{1}, \qquad (4.2)$$

where $\mathbf{1}$ denotes the $|S|$-vector with all components equal to 1. Equation (4.2) follows immediately from (3.20) and (3.21). Thus, by choosing the value of c_k sufficiently large - allowing acceptance of virtually all proposed transitions - quasi equilibrium is directly attained at these values of the control parameter, since in this case all solutions occur with equal probability given by the aforementioned uniform distribution of (4.2). Next, the length of the Markov chains and the decrement function must be chosen such that quasi equilibrium is restored at the end of each individual Markov chain. In this way the equilibrium distributions for the various Markov chains will be 'closely followed', so as to arrive eventually, as $c_k \downarrow 0$, close to $\mathbf{q}^*$, the uniform distribution on the set of optimal solution; see (2.7) or (3.24).

It is intuitively clear that large decrements in c_k will require longer Markov chain lengths in order to restore quasi equilibrium at the next value of the control parameter, c_{k+1}. Thus, there is a trade-off between large decrements of the control parameter and small Markov chain lengths. Usually, one goes for small decrements in c_k to avoid extremely long chains, but alternatively, one could use large values for L_k in order to be able to make large decrements in c_k.

The search for adequate cooling schedules has been the subject of study in many papers published over the years. Reviews are given by Collins, Eglese & Golden [1987] and Van Laarhoven & Aarts [1987]. In the remainder of this section we briefly discuss the original cooling schedule proposed by Kirkpatrick, Gelatt & Vecchi [1982; 1983]. This schedule has been used in many applications of the simulated annealing algorithm and is based on a number of *conceptually simple* empirical rules. A *more elaborate* and theoretically based schedule is discussed in Section 4.2.

The cooling schedule proposed by Kirkpatrick, Gelatt & Vecchi is centered on the parameters of Definition 4.1.

Initial Value of the Control Parameter
As already stated, the value of c_0 should be large enough to allow virtually all transitions to be accepted. This is achieved by requiring that the *initial acceptance ratio* $\chi_0 = \chi(c_0)$ is close to 1. In practice this can be achieved by starting off at a small positive value of c_0 and multiplying it with a constant factor, larger than 1, until the corresponding value of χ_0, calculated from the generated transitions, is close to 1. In the physical analogy discussed in Chapter 2 this corresponds to heating up the solid until all particles are randomly arranged in the liquid phase.

Decrement of the Control Parameter
As mentioned previously, one usually opts for small changes in the value of the control parameter and a frequently used decrement function is given by

$$c_{k+1} = \alpha \cdot c_k, \quad k = 1, 2, \ldots, \tag{4.3}$$

where α is a constant smaller than but close to 1. Typical values lie between 0.8 and 0.99.

Final Value of the Control Parameter
Execution of the algorithm is terminated if the value of the cost function of the solution obtained in the last trial of a Markov chain remains unchanged

for a number of consecutive chains.

Length of the Markov Chains
The length of Markov chains is based on the requirement that at each value c_k of the control parameter quasi equilibrium is to be restored. The number of transitions needed to achieve this is calculated from the intuitive argument that quasi equilibrium will be restored after acceptance of at least some fixed number of transitions. However, since transitions are accepted with decreasing probability, one would obtain $L_k \rightarrow \infty$ for $c_k \downarrow 0$. Consequently, L_k is bounded by some constant $\overline{L}$ to avoid extremely long Markov chains for small values of c_k.

4.2   A Polynomial-Time Cooling Schedule

In this section we discuss a cooling schedule presented by Aarts & Van Laarhoven [1985a; 1985b]. This cooling schedule leads to a polynomial-time execution of the simulated annealing algorithm, but it cannot give any guarantee for the deviation in cost between the final solution obtained by the algorithm and the optimal cost. The discussion presented below is again based on the four parameters constituting the cooling schedule; see Definition 4.1.

Initial Value of the Control Parameter
The initial value c_0 of the control parameter is again obtained from the requirement that at this value of the control parameter virtually all proposed transitions should be accepted. Assume that a sequence of trials is generated at a certain value c of the control parameter. Let m_1 denote the number of proposed transitions from i to j for which $f(j) \leq f(i)$, and m_2 the number of transitions for which $f(j) > f(i)$. Furthermore, let $\overline{\Delta f}^{(+)}$ be the average difference in cost over the m_2 cost-increasing transitions. Then the acceptance ratio χ can be approximated by the following expression:

$$\chi \approx \frac{m_1 + m_2 \cdot \exp\left(\frac{-\overline{\Delta f}^{(+)}}{c}\right)}{m_1 + m_2}, \tag{4.4}$$

from which we obtain:

$$c = \frac{\overline{\Delta f}^{(+)}}{\ln\left(\frac{m_2}{m_2 \cdot \chi - m_1(1-\chi)}\right)}. \tag{4.5}$$

The initial value of c_0 can be calculated from (4.5) in the following way. Initially, c_0 is set equal to zero. Next, a sequence of m_0 trials is generated. After each trial, a new value of c_0 is calculated from (4.5), where χ is set to χ_0; hereafter, χ_0 is referred to as the *initial acceptance ratio*. The values of m_1 and m_2 correspond to the number of cost-decreasing and cost-increasing transitions, respectively, obtained so far; eventually, $m_0 = m_1 + m_2$. Numerical experiments indicate that fast convergence to a final value of c_0 is obtained in this way. This final value is then taken as the initial value of the control parameter.

Decrement of the Control Parameter

In the previous section it was argued that, for small steps in the decrement of the control parameter, the subsequent stationary distributions of the homogeneous Markov chains will be close to each other. As a consequence of this one may expect that, after decreasing c_k to c_{k+1}, a small number of transitions is likely to be sufficient for restoring quasi equilibrium at c_{k+1} provided quasi equilibrium holds at c_k. Next, we assume that the condition for quasi equilibrium given in (4.1) may be replaced by

$$\forall k \geq 0: \ \|\mathbf{q}(c_k) - \mathbf{q}(c_{k+1})\| < \varepsilon, \tag{4.6}$$

for some positive value of ε. So, we assume that quasi equilibrium is maintained throughout the optimization process if (4.6) holds for all k. Evidently, this assumption requires that quasi equilibrium is achieved at c_0. The latter is accounted for by the choice of the initial value of the control parameter.

Thus, for two successive values of the control parameter we want the stationary distributions to be 'close'. This can be quantified by requiring that

$$\forall i \in S: \frac{1}{1+\delta} < \frac{q_i(c_k)}{q_i(c_{k+1})} < 1+\delta, \quad k = 0, 1, \ldots, \tag{4.7}$$

for some small positive number δ, that can be related to ε of (4.6). The following theorem provides a sufficient condition to satisfy (4.7).

Theorem 4.1 *Let $\mathbf{q}(c_k)$ be the stationary distribution for the homogeneous Markov chain associated with the simulated annealing algorithm with components given by (3.20) and (3.21), and let c_k and c_{k+1} be two successive values of the control parameter with $c_{k+1} < c_k$. Then the inequalities of (4.7) are satisfied if the following condition holds:*

$$\forall i \in S: \frac{\exp\left(-\frac{\delta_i}{c_k}\right)}{\exp\left(-\frac{\delta_i}{c_{k+1}}\right)} < 1+\delta, \quad k = 0, 1, \ldots, \tag{4.8}$$

where $\delta_i = f(i) - f_{opt}$.

Proof The components of the stationary distribution given by (3.20) and (3.21) can be rewritten as

$$\forall i \in S : \ q_i(c_k) = \frac{1}{M_0(c_k)} \exp\left(-\frac{\delta_i}{c_k}\right),$$

where the normalization constant M_0 is given by

$$M_0(c_k) = \sum_{j \in S} \exp\left(-\frac{\delta_j}{c_k}\right).$$

Thus, if $c_{k+1} < c_k$, we have

$$\exp\left(-\frac{\delta_i}{c_k}\right) \geq \exp\left(-\frac{\delta_i}{c_{k+1}}\right)$$

and

$$M_0(c_k) \geq M_0(c_{k+1}).$$

Hence, we obtain

$$\forall i \in S : \ q_i(c_k) \ = \ \frac{1}{M_0(c_k)} \exp\left(-\frac{\delta_i}{c_k}\right) \leq \frac{1}{M_0(c_{k+1})} \exp\left(-\frac{\delta_i}{c_k}\right)$$

$$< \ (1+\delta)\frac{1}{M_0(c_{k+1})} \exp\left(-\frac{\delta_i}{c_{k+1}}\right) = (1+\delta)q_i(c_{k+1}).$$

On the other hand we have

$$\forall i \in S : \ q_i(c_k) \ = \ \frac{1}{M_0(c_k)} \exp\left(-\frac{\delta_i}{c_k}\right)$$

$$\geq \ \frac{1}{\sum_{j \in S} \exp\left(-\frac{\delta_j}{c_k}\right)} \exp\left(-\frac{\delta_i}{c_{k+1}}\right)$$

$$> \ \frac{1}{\sum_{j \in S} (1+\delta) \exp\left(-\frac{\delta_j}{c_{k+1}}\right)} \exp\left(-\frac{\delta_i}{c_{k+1}}\right)$$

$$= \ \frac{1}{(1+\delta)} q_i(c_{k+1}),$$

which completes the proof.   ∎

Equation (4.8) can be rewritten to give the following condition on the two subsequent values of the control parameter:

$$\forall i \in S : \ c_{k+1} > \frac{c_k}{1 + \frac{c_k \ln(1+\delta)}{f(i) - f_{opt}}}, \quad k = 0, 1, \dots \tag{4.9}$$

We now introduce a slight simplification by restricting the condition of (4.8) to the set of solutions that occur with most probability during the generation of the kth Markov chain. From Postulate 2.1 we recall that the probability distribution of the cost values of solutions can be reliably approximated by a normal distribution in the region close to the average value of the cost function over all solutions, and by an exponential distribution in the region close to the optimal value of the cost function. Furthermore, from (2.26) and (2.27) we recall that the mean $\langle f \rangle_{c_k}$ and standard deviation σ_{c_k} of these distributions can be approximated by the average cost $\overline{f}(c_k)$ and the spreading $\sigma(c_k)$, respectively. We now define the set S_{c_k} as

$$S_{c_k} = \{i \in S \mid f(i) - f_{opt} \leq \langle f \rangle_{c_k} - f_{opt} + 3\sigma_{c_k}\}. \tag{4.10}$$

Thus, due to the properties of the normal distribution and the exponential distribution, a solution obtained in the kth Markov chain has a probability close to 1 of being in the set S_{c_k}; more precisely, the probability is 0.99 for the normal distribution and 0.95 for the exponential distribution. Using the simplification described above, (4.9) can be replaced by the following condition:

$$c_{k+1} > \frac{c_k}{1 + \frac{c_k \cdot \ln(1+\delta)}{\langle f \rangle_{c_k} - f_{opt} + 3\sigma_{c_k}}}, \quad k = 0, 1, \ldots, \tag{4.11}$$

which is a stronger condition for most solutions generated in the kth Markov chain.

For many instances of combinatorial optimization problems the value of f_{opt} is not known. However, the average value and the spreading of the cost function typically exhibit a similar behaviour as a function of the control parameter; cf. Figure 2.4. Hence, we argue that $\langle f \rangle_{c_k} - f_{opt} + 3\sigma_{c_k}$ can be replaced by $3\sigma_{c_k}$ and that the omission of the term $\langle f \rangle_{c_k} - f_{opt}$ is be counterbalanced by choosing smaller values of δ. Thus, (4.11) can be replaced by the following expression:

$$c_{k+1} = \frac{c_k}{1 + \frac{c_k \cdot \ln(1+\delta)}{3\sigma_{c_k}}}, \quad k = 0, 1, \ldots. \tag{4.12}$$

The amount by which the value of c is decreased by the decrement function of (4.12) is determined by the value of δ, hereafter called the *distance parameter*. Small δ-values lead to small decrements; large δ-values lead to large decrements in c.

Final Value of the Control Parameter

Termination of the algorithm is based on an extrapolation of the expected cost

$\langle f \rangle_{c_k}$ for $c_k \downarrow 0$. Let

$$\Delta \langle f \rangle_c = \langle f \rangle_c - f_{opt}, \qquad (4.13)$$

then execution of the algorithm is terminated if $\Delta \langle f \rangle_c$ is small compared to $\langle f \rangle_{c_0}$, the expected cost at c_0. For sufficiently large values of c_0 we have $\langle f \rangle_{c_0} \approx \langle f \rangle_\infty$; see (2.14). Hence, we may approximate $\Delta \langle f \rangle_c$ for $c \ll 1$ by:

$$\Delta \langle f \rangle_c \approx c \frac{\partial \langle f \rangle_c}{\partial c}. \qquad (4.14)$$

Hence, we may reliably terminate the algorithm if for some k we have

$$\frac{c_k}{\langle f \rangle_\infty} \left. \frac{\partial \langle f \rangle_c}{\partial c} \right|_{c=c_k} < \varepsilon_s, \qquad (4.15)$$

where ε_s is some small positive number. Hereafter we refer to ε_s as the *stop parameter* and to (4.15) as the *stop criterion*.

In practice the value of $\overline{f}(c_k)$, which is used to approximate $\langle f \rangle_{c_k}$, may strongly fluctuate, especially for large values of c_k. This may lead to premature satisfaction of the stop criterion and hence to premature termination of the algorithm. This problem can be circumvented by introducing a smoothing of the value of $\overline{f}(c_k)$. Such a smoothing can be achieved by replacing each data point $(c_k, \overline{f}(c_k))$ by a data point $(c_k, f_s(c_k))$, where $f_s(c_k))$ is the average over a number of consecutive values of $\overline{f}$ around c_k.

Length of the Markov Chains

By using the concept of quasi equilibrium we argue that the decrement function of the control parameter derived above requires only a 'small' number of transitions to rapidly approach the stationary distribution for a given next value of the control parameter. We conjecture that 'small' can be specified as the number of transitions for which the algorithm has a sufficiently large probability of visiting at least a major part of the neighbourhood of a given solution. To quantify this number we need the following theorem.

Theorem 4.2 *Let S denote a set of cardinality S, then the expected fraction of different elements of S that is selected by N random samplings with repetition from S can be approximated, for large N and S, by*

$$1 - e^{-\frac{N}{S}}. \qquad (4.16)$$

Proof The expected fraction of different elements of S, that is selected by N random samplings with repetition from S, equals the probability of selecting a given element from S in N samplings. An expression for this probability can be derived in the following way. The probability of not selecting a given element from the set S in N random samplings is given by $(1 - \frac{1}{S})^N$. Consequently, the probability of selecting the given element from S in N samplings equals

$$\left[1 - \left(\frac{S-1}{S} \right)^N \right].$$

Hence, for $N, S \to \infty$ we obtain

$$\lim_{N \to \infty} \left[1 - \left(\frac{S-1}{S} \right)^N \right] = \lim_{S \to \infty} \left[1 - \left(\frac{S-1}{S} \right)^{xS} \right]$$

$$= 1 - e^{-x},$$

where $x = N/S$. ∎

In our cooling schedule we take the length of the Markov chains equal to the size Θ of the neighbourhoods, i.e.

$$L_k = L = \Theta, \quad k = 0, 1, \dots. \tag{4.17}$$

Now suppose the simulated annealing algorithm starts the generation of a Markov chain of length L with a given solution i, and suppose that no proposed transition is accepted during the generation of the Markov chain. Then, according to (4.16), where $S = \Theta$ and $N = L = \Theta$, the fraction of different neigbouring solutions of i that is visited during the chain-generation process equals $1 - e^{-1} \approx \frac{2}{3}$; for three subsequent Markov chains, this fraction is approximately 1, indicating that approximately all neighbours have been visited. These numbers hold for sufficiently large neighbourhoods, i.e. $\Theta > 100$; cf. Aarts, Beenker & Korst [1985].

Summarizing, we have derived a cooling schedule with three parameters, i.e. the initial acceptance ratio χ_0, the distance parameter δ and the stop parameter ε_s. It should be noted that the parameters do not depend on the problem instance at hand. In Section 4.3 we discuss an empirical study in which we analyse the finite-time behaviour of the simulated annealing algorithm using the cooling schedule presented above. Furthermore, in Section 4.3 we present computational evidence for the assertion that the cooling schedule leads to near-optimal solutions.

We now prove that the cooling schedule presented above leads to a total number of steps in the control parameter bounded by $\mathcal{O}(\ln |S|)$. This is stated by the following theorem.

Theorem 4.3 *Let the decrement function for the control parameter be given by*

$$c_{k+1} = \frac{c_k}{1 + \alpha_k c_k}, \quad k = 0, 1, \ldots \tag{4.18}$$

where

$$\alpha_k = \frac{\ln(1 + \delta)}{3\sigma_{c_k}}, \quad k = 0, 1, \ldots, \tag{4.19}$$

and let K denote the first integer for which the stop criterion is satisfied, i.e.

$$\frac{c_K}{\langle f \rangle_\infty} \frac{\partial \langle f \rangle_c}{\partial c}\bigg|_{c=c_K} < \varepsilon_s. \tag{4.20}$$

Then we have

$$K = \mathcal{O}(\ln |S|), \tag{4.21}$$

under certain conditions on the derivatives (with respect to c) of the expected value $\langle f \rangle_c$ and the entropy S_c.

Proof The proof of the theorem consists of two parts: (i) the total number of steps K is expressed as a function of c_K, the final value of the control parameter, and (ii) a lower bound is derived on c_K. The combination of these two results then yields the proof of the theorem.

(i) It can be straightforwardly shown by induction that

$$c_k \leq \frac{c_0}{1 + k\alpha c_0}, \quad k = 0, 1, \ldots, \tag{4.22}$$

where $\alpha = \min_k \alpha_k$. Indeed, (4.22) holds for $k = 0$. Next, suppose (4.22) holds for k. Then by using $\alpha_k \geq \alpha$ and the induction hypothesis we obtain

$$c_{k+1} = \frac{1}{c_k^{-1} + \alpha_k} \leq \frac{1}{\frac{1+k\alpha c_0}{c_0} + \alpha} = \frac{c_0}{1 + (k+1)\alpha c_0}.$$

Hence, (4.22) holds for $k + 1$. Consequently, we obtain

$$K \leq \frac{c_0 - c_K}{\alpha c_0 c_K} < \frac{1}{\alpha c_K} = \frac{3 \max_k \sigma_{c_k}}{\ln(1+\delta)c_K} \approx \frac{3\sigma_\infty}{\ln(1+\delta)c_K},$$

where σ_∞ is given by (2.16).

(ii) If quasi equilibrium is restored at each value of c_k we may assume that the equilibrium conditions discussed in Section 2.3 hold and that we may use the following relation; cf. (2.12) and (2.13):

$$\frac{\partial}{\partial c}\langle f\rangle_c = c\frac{\partial}{\partial c}S_c.$$

(4.23)

From (2.13) we recall that S_c is increasing with c. Furthermore, from (2.18) and (2.19) we recall that

$$S_\infty = \ln|S|,$$

(4.24)

and

$$S_0 = \ln|S_{opt}|$$

(4.25)

Now, using (4.23)-(4.25), K can be expressed in terms of $\ln|S|$ as follows. From (4.20) we conclude that

$$\exists \varepsilon' \in (0, \varepsilon_s] : \ \varepsilon' = \frac{c_K}{\langle f\rangle_\infty}\frac{\partial\langle f\rangle_c}{\partial c}\bigg|_{c=c_K}.$$

Hence, using (4.23), we obtain

$$\varepsilon' = \frac{c_K^2}{\langle f\rangle_\infty}\frac{\partial S_c}{\partial c}\bigg|_{c=c_K}.$$

Since $c_K \ll 1$, we may approximate $\frac{\partial}{\partial c}S_c\big|_{c=c_K}$ by $\frac{S_K - S_0}{c_K}$, which yields the following relation:

$$\varepsilon' \approx \frac{c_K^2}{\langle f\rangle_\infty}\frac{S_K - \ln|S_{opt}|}{c_K} < \frac{c_K\ln|S|}{\langle f\rangle_\infty}.$$

Consequently,

$$c_K > \frac{\langle f\rangle_\infty\varepsilon'}{\ln|S|}$$

and

$$K < \frac{\ln|S|}{\alpha\langle f\rangle_\infty\varepsilon'} \approx \frac{3\sigma_\infty\ln|S|}{\ln(1+\delta)\langle f\rangle_\infty\varepsilon'}.$$

(4.26)

Thus, the total number of steps in the control parameter is bounded by $\eta\ln|S|$, where $\eta = \frac{3\sigma_\infty}{\ln(1+\delta)\langle f\rangle_\infty\varepsilon'}$. ∎

We finally conclude that, if the approximations used in the proof are sufficiently accurate, the simulated annealing algorithm, using the three-parameter schedule presented above, requires a computation time T that satisfies the following relation:

$$T = O(\tau \cdot L \cdot \ln |S|), \tag{4.27}$$

where L denotes the length of the individual Markov chains, $\ln |S|$ denotes the upper bound on the number of Markov chains, and τ denotes the computation time required to carry out a single transition. For most combinatorial optimization problems, L and τ can be chosen polynomially in the problem size. Consequently, if $\ln |S|$ is polynomial in the problem size, which holds for many combinatorial optimization problems, the simulated algorithm algorithm runs in polynomial time.

4.3   Empirical Performance Analysis

In the previous section we presented a polynomial-time cooling schedule using three parameters: the initial acceptance ratio χ_0, the distance parameter δ, and the stop parameter ε_s. In this section we discuss the finite-time behaviour of the simulated annealing algorithm by means of an empirical performance analysis. The analysis is based on the polynomial-time cooling schedule discussed in the previous section investigating the performance of the simulated annealing algorithm as a function of the three parameters mentioned above.

Traditionally, the performance of an approximation algorithm is related to the following two quantities:

  - the *quality* of the final solution obtained by the algorithm, and

  - the *running time* required by the algorithm.

The quality of a final solution with cost f^* can be quantified by the *error* $\mathcal{E}$ which is defined as

$$\mathcal{E} = \frac{f^* - f_{opt}}{f_{opt}}, \tag{4.28}$$

i.e. $\mathcal{E}$ is the relative difference between the final cost f^* and the optimal cost f_{opt}. The running time can be quantified by the number of elementary[1] transitions or by the CPU time required by the algorithm.

The performance of the simulated annealing algorithm has been investigated by carrying out an average-case analysis. *Average case* relates here to

---

[1] For instance, in the case of the simulated annealing algorithm, the number of proposed transitions would be a good measure.

the average value of the error and the running time computed from the probability distribution over the set of final solutions that can be obtained by the algorithm for a given problem instance.[2] This probability distribution results from the probabilistic nature of the simulated annealing algorithm.

In our investigations we applied the simulated annealing algorithm to the EUR100 instance of the travelling salesman problem introduced in Chapter 2; see also the Appendix. Average values of the error and the running time were obtained by running the simulated annealing algorithm a number of times, between 5 and 10, for the same problem instance using different initial solutions.

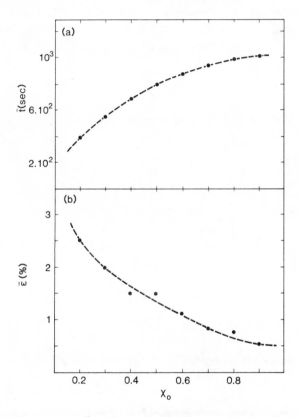

Figure 4.1: (a) Average running time and (b) average error as a function of χ_0 for fixed values of δ and ε_s (problem instance: EUR100, $\delta = 0.1$, $\varepsilon_s = 10^{-5}$).

Figure 4.1 shows (a) the average running time and (b) the average error as a function of the initial acceptance ratio χ_0. The values of the distance param-

---

[2]In the analysis of deterministic algorithms, average case refers to a probability distribution over a set of problem instances.

eter δ and the stop parameter ε_s are fixed. Figure 4.1a shows that starting off at lower values of the control parameter, i.e. using smaller values of χ_0, leads to faster execution of the algorithm. Clearly, this is what one would expect since the algorithm starts off at smaller values of the control parameter, while it terminates at approximately the same value of the control parameter.

From Figure 4.1b it is observed that the average error increases as χ_0 decreases. This behaviour can be explained as follows. Smaller values of the initial acceptance ratio result in lower values of the initial control parameter; see (4.5). Consequently, for these smaller values, the initial condition for quasi equilibrium, i.e. large initial value of the control parameter, is no longer met. This obviously results in a deterioration of the quality of the final solution, i.e. an increase of the error, since the equilibrium distribution is no longer closely followed.

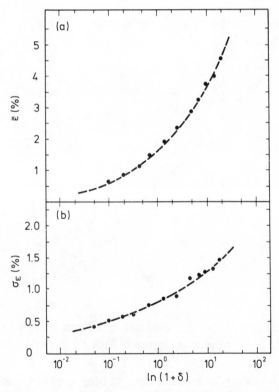

Figure 4.2: (a) Average error and (b) spreading of the error as a function of the distance parameter δ for fixed values of χ_0 and ε_s (problem instance: EUR100, $\chi_0 = 0.95$, $\varepsilon_s = 10^{-5}$).

Figure 4.2 shows (a) the average error and (b) the spreading of the error as a function of the distance parameter δ; the values of χ_0 and ε_s are fixed. The figures show that, as δ decreases, both the average error and the spreading of the error decrease. From the definition of quasi equilibrium we know that, for smaller values of δ, the stationary distributions will be followed more closely with decreasing values of the control parameter. Thus, the probability of obtaining a high-quality final solution will increase as δ decreases. This behaviour is clearly reflected in Figure 4.2.

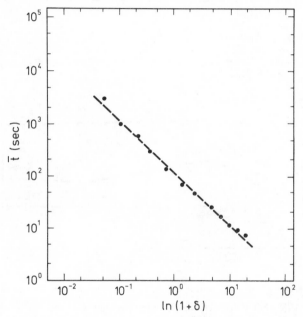

Figure 4.3: Average running time as a function of the distance parameter δ for fixed values of χ_0 and ε_s (problem instance: EUR100, $\chi_0 = 0.95$, $\varepsilon_s = 10^{-5}$).

The behaviour shown in Figure 4.2 can be parametrized by the following expressions:

$$\overline{\mathcal{E}} \quad \propto \quad (\ln(1+\delta))^a \text{ and} \tag{4.29}$$

$$\sigma_{\mathcal{E}} \quad \propto \quad (\ln(1+\delta))^{\frac{a}{2}}, \tag{4.30}$$

where a denotes some positive constant; here $a = 0.42$, $\chi^2 = 0.98$. The typical behaviour of the average error and its spreading $(\overline{\mathcal{E}} \approx \sigma_{\mathcal{E}}^2)$ can be explained by assuming that the values of f^*, the costs of the final solutions, are distributed according to a gamma distribution of the form

$$\mathcal{G}_p(x) = \frac{1}{\Gamma(p)} x^{p-1} e^{-x}, \qquad (4.31)$$

where $x = f^* - f_{opt}$ and $p \propto (\ln(1 + \delta))^a$ [Aarts, Korst & Van Laarhoven, 1988]. Furthermore, defining the *cooling rate* as the average decrement in the control parameter per trial, it follows immediately that the cooling rate is proportional to $\ln(1 + \delta)$ and that the average error is proportional to the ath power of the cooling rate.

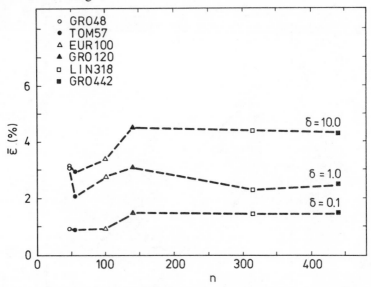

Figure 4.4: Average error as a function of the problem size in the travelling salesman problem for different values of the distance parameter δ and for fixed values of χ_0 and ε_s ($\chi_0 = 0.95$, $\varepsilon_s = 10^{-5}$). The data are taken from Van Laarhoven [1988]; see also Aarts & Van Laarhoven [1988].

Figure 4.3 shows the average running time as a function of δ; the values of χ_0 and ε_s are fixed. Evidently, smaller values of δ require larger amounts of computation time. This is a direct consequence of the nature of the decrement function of (4.12), viz. smaller values of δ lead to more decrement steps.

Figure 4.4 shows the average error for a number of differently sized instances of the travelling salesman problem for three different values of the distance parameter δ; the values of χ_0 and ε_s are fixed. The instances are taken from the literature and can be divided into the following two sets.

- Euclidean problem instances: GRO48 and GRO120 are instances with 48 and 120 German cities, respectively, and are due to Grötschel [1977].

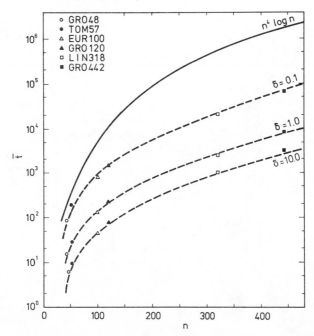

Figure 4.5: Average running time as a function of the problem size in the travelling salesman problem for different values of the distance parameter δ and for fixed values of χ_0 and ε_s ($\chi_0 = 0.95$, $\varepsilon_s = 10^{-5}$). The data are taken from Van Laarhoven [1988]; see also Aarts & Van Laarhoven [1988].

TOM57 is an instance with 57 cities in the United States and is due to Karg & Thompson [1964]. EUR100 is again the instance with the 100 European cities due to Aarts, Korst & Van Laarhoven [1988]; see also the Appendix.

- Non-Euclidean problem instances: LIN318 is a 318-city problem instance due to Lin and Kernighan [1973], and GRO442 is a 442-city problem instance due to Grötschel [1984]. Both instances originate from the problem of finding the shortest route for a drilling machine through a number of points. Instance GRO442 is one of the largest instances of the travelling salesman problem reported in the literature that has been solved to optimality without using partitioning techniques.

From the results presented in Figure 4.4 we conjecture that the error of the final solutions is independent of the problem size, when applying the polynomial cooling schedule of Section 4.2, for a fixed set of parameter values.

Figure 4.5 shows the average running time for different-sized instances of the TSP, and for different values of the distance parameter; the values of χ_0

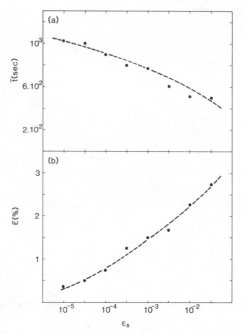

Figure 4.6: (a) Average running time and (b) the average error as a function of the stop parameter ε_s for fixed values of χ_0 and δ (problem instance: EUR100, $\chi_0 = 0.95$, $\delta = 0.1$).

and ε_s are fixed. It is observed that the average-case time complexity is the same for the different δ-values (dashed lines), and is estimated to be slightly less than $\mathcal{O}(n^3 \ln n)$, where n denotes the number of cities. According to (4.27) the worst-case time complexity is $\mathcal{O}(n^4 \ln n)$, since $\Theta = (n - 1)(n - 2)$, $|S| = (n - 1)!$ and $\tau = \mathcal{O}(n)$; see Example 1.3. From Figure 4.5 we observe that the average-case time complexity is close to the worst-case time complexity (solid line). Finally, it should be noted that running times may be very large at small values of δ for the larger problem instances.

Figure 4.6 shows (a) the average running time and (b) the average error as a function of the stop parameter ε_s; the values of χ_0 and δ are fixed. The observed behaviour is evident, viz. as the value of ε_s increases the stop criterion of (4.15) is more easily met and, consequently, the running time decreases, while the error increases.

From the analysis presented above we conclude that the effect of variations in the values of the parameters χ_0, δ and ε_s on the performance of the simulated annealing algorithm is large for the distance parameter δ and only minor for the initial acceptance ratio χ_0 and the stop parameter and ε_s. Furthermore, we mention that the behaviour discussed above is observed for a number of other problems to which the simulated annealing algorithm has been applied [Aarts & Van Laarhoven, 1985a; 1988; Van Laarhoven, Aarts & Lenstra, 1988; Van Laarhoven, 1988].

We thus arrive at the following conclusions on the performance of the simulated annealing algorithm using the polynomial-time cooling schedule presented in Section 4.2.

- The algorithm has the potential to find high-quality solutions but at the cost of substantial computational efforts.

- The algorithm is robust, i.e. its final solution does not strongly depend on the choice of the initial solution. The spreading of the error is small for a set of fixed parameters in the cooling schedule.

- The quality of the final solution is predominantly dependent on the distance parameter which determines the speed at which the control parameter is lowered. The values of the parameters determining the initial and final values of the control parameter play only a minor role as long as they are chosen with a reasonable extent of accuracy.

- The average-case running time is close to the worst-case running time.

In the next chapter we present a survey of problems and problem areas to which the simulated annealing algorithm has been applied. Furthermore, we discuss some performance characteristics. We can already mention that the characteristic features listed above are also reported by other authors using different cooling schedules. An important conclusion from this is that cooling schedules do not strongly influence the final results, both in quality and running time, as long as the cooling is done accurately; cf. Aarts & Van Laarhoven [1988], Johnson, Aragon, McGeoch & Schevon [1988], and Van Laarhoven [1988]. Unfortunately, not every cooling schedule reported in the literature turns out to deserve the label 'accurately'; cf. Van Laarhoven & Aarts [1987]. Thus, choosing a cooling schedule should be done carefully.

CHAPTER 5

Simulated Annealing in Practice

In this chapter we discuss a number of general aspects related to the application of the simulated annealing algorithm such as the problem representation and the transition mechanism. Using these aspects as a guideline, we discuss a number of combinatorial optimization problems as examples of problems to which the simulated annealing algorithm can be applied. The examples discussed here are the *travelling salesman problem*, the *max cut problem*, the *independent set problem*, the *graph colouring problem*, and *the placement problem*. These problems are chosen so as to illustrate a number of characteristic features of the application of the simulated annealing algorithm, such as the problem formulation, incremental calculations of the cost differences, the use of penalty functions and the transformation of the problem at hand to an equivalent problem to which the algorithm can be applied more efficiently.

Furthermore, we present in this chapter a survey of problems in different problem areas to which the simulated annealing algorithm has been applied. The chapter is concluded with a discussion of some general performance experiences obtained with the algorithm.

5.1 Implementing the Algorithm

In applying the simulated annealing algorithm one commonly resorts to an implementation in which a sequence of Markov chains is generated at descending values of the control parameter; see Chapter 4. Individual Markov chains are generated by continuously trying to transform a current solution into a subsequent one by applying a generation mechanism and an acceptance criterion. Application of the simulated annealing algorithm requires specification of three distinct items: (i) a concise problem representation, (ii) a transition mechanism, and (iii) a cooling schedule. We shall now elaborate on these

77

items in more detail.

(*i*) A concise description of the *problem representation* consists of a representation of the solution space and an expression of the cost function. The cost function must be chosen such that it represents the cost effectiveness of the solutions with respect to the optimization objective. Both the problem representation and the cost function should be given by simple expressions that are easy to manipulate.

(*ii*) Generating trails for transforming a current solution into a subsequent one consist of three steps. Firstly, a new solution is generated from a current one by applying a generation mechanism. Secondly, the difference in cost between the two solutions is calculated, and thirdly, a decision is made on whether or not the new solution is to be accepted.

The evaluation of trails is the most time consuming part of the simulated annealing algorithm and therefore should be done as time efficiently as possible. The generation mechanism is usually chosen such that new solutions are obtained from current ones by simple rearrangements that can be computed rapidly, for instance permutations, swapping, and inversions. Calculation of cost differences is preferably done incrementally, taking into account only the differences resulting from the local rearrangements. For most applications this is the fastest way to calculate the differences in cost.

The decision to accept new solutions is based on the acceptance criterion. Most frequently one applies the Metropolis criterion which is given by

$$\mathbb{P}\{accept\} = \begin{cases} 1 & \text{if } \Delta f < 0 \\ \exp\left(-\frac{\Delta f}{c}\right) & \text{if } \Delta f \geq 0, \end{cases} \qquad (5.1)$$

where c denotes the control parameter and Δf the difference in cost between a new and a current solution in the case of a minimization problem and *vice versa* in the case of a maximization problem; see also (2.4) and (3.8).

(*iii*) Carrying out the optimization along the lines of an annealing process requires specification of the parameters determining the cooling schedule; see Chapter 4. These parameters are the initial value of the control parameter, a decrement function of the control parameter, the length of the individual Markov chains and a stop criterion. For instance, applying the polynomial-time cooling schedule presented in Section 4.2, requires the specification of the initial acceptance ratio χ_0, the distance parameter δ, and the stop parameter ε_s.

Below, we discuss the application of the simulated annealing algorithm to a number of combinatorial optimization problems. The discussion is centred on the different aspects related to items (i) and (ii). The aspects related to item (iii) have been extensively discussed in Chapter 4.

The neighbourhoods for the various problems, presented below, are chosen such that the corresponding generation mechanisms induce irreducible and aperiodic Markov chains, thus guaranteeing asymptotic convergence to the set of optimal solutions; see also Section 3.2. The corresponding proofs are straightforward and left to the reader as an exercise.

Finally, we mention that all the problems treated here belong to the class of NP-complete problems [Garey & Johnson, 1979].

5.1.1 The Travelling Salesman Problem

From Examples 1.1 and 1.3 we recall the following.

Definition 5.1 (Travelling salesman problem) Let n be the number of cities and $D = [d_{ij}]$ be the distance matrix whose elements d_{ij} denote the distance between city i and city j. The problem then is to find the shortest tour visiting all cities exactly once. ∎

The simulated annealing algorithm then can be applied in the following way.

- The solution space S is represented by the set of all cyclic permutations $\pi = (\pi(1), \ldots, \pi(n))$, where $\pi(i)$, $i = 1, \ldots, n$, denotes the successor city of city i in the tour represented by π.

- The cost function, which is to be minimized, is chosen as

$$f(\pi) = \sum_{i=1}^{n} d_{i,\pi(i)}. \qquad (5.2)$$

- New solutions can be generated by choosing two arbitrary cities p and q, and reversing the sequence in which the cities in between cities p and q are traversed, i.e. the 2-change generation mechanism of Example 1.3.

- The difference in cost can be calculated incrementally from the following expression; see Figure 1.1:

$$\Delta f = -d_{p,\pi(p)} - d_{\pi^{-1}(q),q} + d_{p,\pi^{-1}(q)} + d_{\pi(p),q}. \tag{5.3}$$

5.1.2 The Max Cut Problem

Graph partitioning problems - min cut or max cut partitioning, either weighted or not - constitute a large class of combinatorial optimization problems [Garey & Johnson, 1979] with great practical relevance [Ullman, 1984]. Here we treat the max cut problem as an example.

Definition 5.2 (Max cut problem) Given a graph $G = (V, E)$ with positive weights on the edges, find a partition of V into disjoint sets V_0 and V_1 such that the sum of the weights of the edges from E that have one endpoint in V_0 and one endpoint in V_1 is maximal.   ∎

The simulated annealing algorithm can be applied in the following way.

- The solution space consists of all possible partitions of the set V into two sets V_0 and V_1.

- The cost function, which is to be maximized, is chosen as

$$f(V_0, V_1) = \sum_{\{u,v\} \in \delta(V_0, V_1)} w(\{u, v\}), \tag{5.4}$$

where $w(\{u, v\})$ denotes the weight of edge $\{u, v\} \in E$, and $\delta(V_0, V_1)$ the *cut* of a partition of V into V_0 and V_1, which is defined as

$$\delta(V_0, V_1) = \{\{u, v\} \in E \mid u \in V_0 \wedge v \in V_1\}. \tag{5.5}$$

- New solutions are generated by randomly choosing a vertex $u' \in V$ and moving it from V_0 to V_1 if $u' \in V_0$, or *vice versa* if otherwise. The difference in cost then is given by

$$\Delta f = \left(\sum_{\{u',v\} \in E \setminus \delta(V_0, V_1)} w(\{u', v\}) - \sum_{\{u',v\} \in \delta(V_0, V_1)} w(\{u', v\}) \right). \tag{5.6}$$

5.1.3   The Independent Set Problem

In many combinatorial optimization problems we have, in addition to a cost function, a set of constrains that must be satisfied by a particular solution; see for instance Definition 5.4. For such problems it is often appropriate to partition the set of solutions S into two subsets, viz. the set of feasible solutions, i.e. solutions satisfying the constraints, and the set of infeasible solutions, i.e. solutions not satisfying the constraints. Hence, Definition 1.2 can be extended as given below.

Definition 5.3 An instance of a combinatorial optimization problem can be formalized as a triple (S, S', f), where S denotes the set of all possible solutions, both feasible and infeasible ones, S' the set of feasible solutions, and f the cost function defined as a mapping

$$f : S \rightarrow \mathbb{R}. \tag{5.7}$$

Now - in the case of minimization - the problem is to find an optimal solution $i_{opt} \in S'$ such that

$$f(i_{opt}) \leq f(i), \quad \text{for all } i \in S'. \tag{5.8}$$

Clearly, if $f(i_{opt}) < f(i)$, for all $i \in S \backslash S'$, then the problem in Definition 5.3 is equivalent to the problem in Definition 1.2. ■

Infeasibility of solutions can be used advantageously by the simulated annealing algorithm. As an example of this we discuss the following problem.

Definition 5.4 (Independent set problem) Given a graph $G = (V, E)$, find the largest independent set, i.e. find the largest set $V' \subset V$, such that for all $u, v \in V'$ the edge $\{u, v\}$ is not in E. ■

The simulated annealing algorithm then can be applied in the following way.

- The solution space S consists of all possible partitions of the set V into sets V' and $V \backslash V'$. The set S' then consists of all partitions where $\forall u, v \in V' : \{u, v\} \notin E$.

- The cost function, which is to be maximized, is chosen as

$$f(V') = |V'| - \lambda |E'|, \tag{5.9}$$

where E' denotes the set of edges $\{u, v\} \in E$ with $u, v \in V'$, and λ denotes a *weighting factor* which must be larger than 1. Feasible

solutions will contribute only to the first term of the cost function. Infeasible solutions will also contribute to the second term and thus lower
the cost for partitions with equal cardinality. Hence, the second term
can be viewed as a *penalty function*; it penalizes the presence of edges
between the subsets of the partition.

- New solutions are generated by randomly choosing a vertex $u' \in V$
  and moving it from V' to $V \setminus V'$ if $u' \in V'$, or *vice versa* if otherwise.
  The difference in cost then is given by

$$\Delta f = (\chi_{(V \setminus V')}(u') - \chi_{(V')}(u')) \left(1 - \lambda \sum_{\{u',v\} \in E, v \in V'} 1 \right). \qquad (5.10)$$

Clearly, the solution space for the independent set problem could have been
restricted to the set of feasible solutions only. However, a mechanism generating a new feasible solution from a current one would be quite complicated and
time consuming. Furthermore, the presence of the infeasible solutions leads
to a smoothing of the cost function 'landscape', which enables the simulated
annealing algorithm to escape more easily from local optima: there are more
solutions with small cost differences in the vicinity of a local maximum, thus
increasing the escape probability. On the other hand, the sizes of the solution
space and the neighbourhoods increase, resulting in longer Markov chains for
most cooling schedules. However, for a number of problems, enlarging the
set of feasible solutions with the set of infeasible solutions may enable the use
of simpler generation mechanisms which often leads to faster convergence.

5.1.4   The Graph Colouring Problem

Definition 5.5 (Graph colouring problem) Given a graph $G = (V, E)$, find
a minimal colouring, i.e. find a mapping $f : V \to \{1, 2, \ldots, k\}$ such that
$f(u) \neq f(v)$, for all $u, v \in V$ with $\{u, v\} \in E$, and k is minimal.   ∎

The graph colouring problem can be viewed as the problem of finding a partition of the set of vertices into a minimal number of independent sets. Such a
partition is called a *colouring*. In constructing an appropriate solution space
we can, similarly to the independent set problem, either use a solution space
consisting of all possible colourings, i.e. the set of feasible solutions, or use a
solution space consisting of all possible partitions. Obviously, partitions not
corresponding to colourings are infeasible.

In the case of a solution space consisting of feasible solutions only, the generation mechanism must ensure that newly generated solutions are also feasible. This can be done by using the concept of *Kempe chains* as was shown by Morgenstern & Shapiro [1986]. The approach pursued by these authors can be described as follows. Let V_i and V_j be two sets of the partition of V and let $v \in V_i$. Then the Kempe chain $i - j$, associated with v, is given by the connected subgraph $G_{(v)} \subset G$ obtained by searching outwards from v, traversing only those edges in E connecting vertices either in V_i or V_j. Changing v from V_i to V_j does not affect feasibility if all vertices in $G_{(v)}$ are changed either from V_i to V_j or *vice versa*.

As in the independent set problem, we argue that such a generation mechanism is time consuming and that it is more convenient, also for reasons of convergence, to enlarge the solution space by adding to the set of feasible solutions the set of infeasible solutions that correspond to a partition but not to a colouring.

Following the independent set problem, a straightforward choice of the cost of a partition in the graph colouring problem is given by a weighted sum of the number of sets in the partition and a penalty term, which penalizes the presence of edges between the subsets of the partition. However, a cost function using only the number of sets cannot discriminate between partitions with different cardinalities of the subsets in the partition. Thus, the cost function 'landscape' will contain large plateaus separated by steep walls. Consequently, there will be no 'guidance' towards local maxima, resulting in a slow convergence of the simulated annealing algorithm. This problem can be dissolved by transforming the graph colouring problem into a problem that is equivalent, in the sense that solving the problem leads to the same set of solutions as in the graph colouring problem, but which allows faster convergence of the simulated annealing algorithm. This can be done in the following way.

Given an instance of the graph colouring problem, then it is easily verified that the number of colours necessary to colour the graph is bounded by $\Delta + 1$, where Δ denotes the maximum degree of the graph [Harary, 1972]. Next, we introduce a set of positive weights $\{w_1, \ldots, w_{\Delta+1}\}$ and a cost function, such that, for a given colouring of G, each vertex contributes w_i to the cost function if it is given colour i. To be able to discriminate between colourings that use a different number of colours, different colours are assigned different weights, such that the cost function is large whenever few colours - each colour with a large weight - are used to colour G. This leads to the following equivalence.

Theorem 5.1 *Let* $W = \{w_1, \ldots, w_{\Delta+1}\}$ *be a set of positive weights satisfying* *the following recursive relation for $j = 1, \ldots, \Delta$:*

$$w_{j+1} \;<\; \left(\frac{|V|}{j} - 1\right) \sum_{i=2}^{j} w_i - \left(\frac{|V|(j-1)}{j} - j\right) w_1. \qquad (5.11)$$

Then solving the graph colouring problem is equivalent to finding a partition $l = \{V_1, \ldots, V_{\Delta+1}\}$ of V such that

$$\forall v_i, v_j \in V_k, \; k = 1, \ldots, \Delta+1 \; : \; \{v_i, v_j\} \notin E, \qquad (5.12)$$

and

$$f(l) = \sum_{i=1}^{\Delta+1} w_i |V_i| \qquad (5.13)$$

is maximal.

Proof Let i denote a feasible partition, i.e. a colouring, of V into the $\Delta + 1$ subsets, and $n(i)$ the number of non-empty subsets in the partition. Clearly, $n(i)$ is identical to the number of colours in i. Then it suffices to show that for two arbitrary colourings i and j the following relation holds:

$$n(i) < n(j) \Rightarrow f(i) > f(j) \qquad (5.14)$$

If (5.14) holds, then the cost function is maximal for a colouring that uses a minimal number of colours.

Let $n_l(i)$ denote the number of vertices of colour l in colouring i, then $\sum_{j=1}^{\Delta+1} n_j(i) = |V|$ for every colouring i. Without loss of generality we may put

$$w_1 > w_2 > \ldots > w_{\Delta+1}.$$

Since we may restrict ourselves to colourings for which (5.13) is maximal over the set of colourings that can be obtained from colouring i by permuting the complete set of colours, we have

$$n_1(i) \geq n_2(i) \geq \ldots \geq n_{\Delta+1}(i).$$

Consequently, $f(i)$ can be written as $\sum_{j=1}^{\Delta+1} w_j n_j(i)$. Now, we must show that (5.11) provides a sufficient condition to ensure that $f(i) > f(j)$ for any two colourings i and j with $n(i) = k$ and $n(j) = k + 1$.

For a colouring i with $n(i) = k$ the lower bound on $f(i)$ is given by

$$f(i) \geq \frac{|V|}{k} \sum_{j=1}^{k} w_j. \qquad (5.15)$$

This follows immediately from *Chebyshev's inequality*, which states that

$$n \sum_{i=1}^{n} a_i b_i \geq \sum_{i=1}^{n} a_i \sum_{i=1}^{n} b_i$$

whenever

$$a_1 \geq a_2 \geq \cdots \geq a_n \text{ and } b_1 \geq b_2 \geq \cdots \geq b_n.$$

For a colouring j, with $n(j) = k + 1$, the upper bound on $f(j)$ is given by

$$f(j) \leq (|V| - k)w_1 + \sum_{i=2}^{k+1} w_i. \tag{5.16}$$

The upper bound is attained when all vertices are given colour 1, except for k vertices which are each given a colour other than 1.

From (5.15) and (5.16) it follows directly that $f(i) > f(j)$ if the weights w_i satisfy (5.11). Hence, if the constraints of (5.11) and (5.12) are satisfied, then the objective function of (5.13) is maximal for a partition that corresponds to a minimal colouring.   ∎

Construction of the value of the weights $w_1, \ldots, w_{\Delta+1}$ for a given Δ such that the constraints of (5.11) are satisfied is straightforward. However, these values are restrictive in the sense that they may become very large for large graphs leading to slow convergence of the simulated annealing algorithm [Korst & Aarts, 1988].

The constraints of (5.11) may be relaxed by choosing the weights according to

$$w_{j+1} < 2w_j - w_1, \quad j = 1, \ldots, \Delta. \tag{5.17}$$

For every Δ it is possible to find positive weights, even integer ones, that satisfy these constraints. If, for instance, a maximum of seven colours is necessary, then the weights w_1 through w_7 can be chosen as 100, 99, 97, 93, 85, 69 and 37, respectively.

Using the constraints of (5.17) results in a much faster convergence of the algorithm. However, in this case, a solution for which (5.13) is maximal does not necessarily correspond to a minimal colouring. Practice shows that these situations hardly occur and moreover, if they do occur, the constraints of (5.17) still result in a near-optimal colouring [Korst & Aarts, 1988].

The simulated annealing algorithm now can be applied to the graph colouring problem in the following way.

- The solution space consists of all possible partitions of the set of vertices V into the sets $V_1, V_2, \ldots, V_{\Delta+1}$, where Δ denotes the maximal degree of the graph G.

- The cost function, which is to be maximized, is chosen as

$$f(V_1, \ldots, V_{\Delta+1}) = \sum_{i=1}^{\Delta+1} w_i(|V_i| - \lambda|E_i|), \qquad (5.18)$$

where w_i denotes the weight assigned to colour i, given by (5.17), E_i denotes the set of edges $\{u, v\} \in E$ with $u, v \in V_i$, and λ denotes a weighting factor which should be larger than 1. As in the independent set problem, the second term in the cost function is again a penalty function. Feasible solutions will only contribute to the first term and infeasible solutions will be penalized by the second term.

- New solutions are generated by randomly choosing a vertex $u' \in V$ and moving it from the current subset to one of the others. Let $u' \in V_k$ be moved to V_l, then the difference in cost is given by

$$\Delta f = w_l - w_k - \lambda \left(w_l \sum_{\{u',v\} \in E, v \in V_l} 1 - w_k \sum_{\{u',v\} \in E, v \in V_k} 1 \right). \quad (5.19)$$

5.1.5   The Placement Problem

Placement problems are well-known in the field of *facilities layout* [Foulds, 1983] and *VLSI layout* [Soukup, 1981]. They typically constitute a class of so-called 'dirty' practical problems for which it is very hard to construct efficient and effective approximation algorithms. The literature presents a large variety of placement problems, all of them being different with respect to additional constraints or optimization objectives, e.g. placement of identical blocks, *gate array placement*, or of shapeable blocks, *floorplanning*; see also the references listed in the item 'placement problems' of Section 5.2.2.

Here, we confine ourselves to the central problem which can be defined as follows.

Definition 5.6 (Placement problem) Given a set of n rectangular blocks and a set of weights $w_{ij}, i, j = 1, \ldots, n$. The problem then is to find a *placement*, i.e. an assignment of the blocks to points of a rectangular grid, such that the blocks do not overlap, and the cost function given by the weighted sum

$$f = A + \lambda C \qquad (5.20)$$

is minimal, where A denotes the area of the rectangle enveloping all blocks, and C a connectivity term given by

$$\sum_{i=1}^{n} \sum_{j=i+1}^{n} w_{ij} d_{ij}, \qquad (5.21)$$

where d_{ij} denotes the distance between two blocks i and j in a given placement. λ denotes a positive weighting factor. ∎

The weights w_{ij} may correspond to adjacency requirements in a facilities layout problem or to connectivity properties in a VLSI layout problem.

The placement problem is NP-complete. A reduction from the quadratic assignment problem is straightforward.

Since the simulated annealing algorithm is very flexible, it is an attractive candidate for application to the placement problem. This can be done in the following way.

- The solution space is chosen as the set of all placements, and thus consists of both feasible solutions, i.e. placements with no overlap, and infeasible solutions, i.e. placements with overlap.

- The cost function, which is to be minimized, usually contains three terms, i.e.

$$f = \lambda_A f_A + \lambda_W f_W + \lambda_O f_O, \qquad (5.22)$$

where f_A denotes the area of the enveloping rectangle, f_W the weighted sum of (5.21), and f_O the amount of overlap in the given placement, which is used as a penalty function. The constants λ_A, λ_W and λ_O are positive factors determining the relative weights of the three different terms in the cost function. Clearly, overlap should be the most strongly penalized. To ascertain feasibility of final solutions, the overlap function is often divided by the control parameter c, i.e. the term $\lambda_O f_O$ is replaced by $\lambda_O f_O / c$. The effect of this is obvious: since the value of c is decreased monotonically in the course of the simulated annealing algorithm, the penalty for overlap increases as c becomes smaller. Hence, overlap is likely to be entirely removed as the algorithm proceeds. Moreover, as for the independent set and the graph colouring problems, allowing infeasibility (overlap) results in a better performance of the simulated annealing algorithm: fast convergence to high-quality solutions.

- New solutions are generated by assigning a subset of the blocks to new grid points. This can be done by local rearrangements of a single block: translation, rotation or inversion, or by swapping two blocks. Clearly, there are many possible variations on this theme. In most implementations of the simulated annealing algorithm for the placement problem, cost differences are calculated incrementally using special data structures such as linked lists. These typical engineering aspects are considered beyond the scope of this book. Examples can be found in the literature, e.g. see De Bont, Aarts, Meehan & O'Brien [1988] and Sechen & Sangiovanni-Vincentelli [1985]; see also Section 5.2.2.

In this section we discussed the application of the simulated annealing algorithm to a number of combinatorial optimization problems. Evidently, this is only a small subset of problems to which the algorithm can be applied. To demonstrate the general applicability of the algorithm and to encourage further reading, we present in the next section a survey of problems to which the algorithm has been applied.

5.2   A Survey of Applications

Ever since its introduction in 1983, the simulated annealing algorithm has been applied to a large number of different combinatorial optimization problems in areas as diverse as operations research, VLSI design, code design, image processing, molecular physics. In many of these areas the existing algorithms performed poorly and therefore the time was ripe for a new algorithm with such salient features as general applicability, flexibility, ease of implementation, and a modicum of sophistication.

Starting off with the seminal paper by Kirkpatrick, Gelatt & Vecchi [1983], reviews of the application of the simulated annealing algorithm have been given by a number of authors. We mention the books by Van Laarhoven & Aarts [1987] and Wong, Leong & Liu [1988], the bibliographies by Collins, Eglese & Golden [1987] and Wille [1986a], and the review papers by Aarts & Van Laarhoven [1987] and Hajek [1985]. For those interested, we mention that the relation between statistical physics and simulated annealing is (extensively) treated by Fu & Anderson [1986], Khachaturyan [1986], Kirkpatrick & Toulouse [1985], Mézard [1987], Mézard & Parisi [1985], and Morgenstern [1987].

In the remainder of this section we list a number of applications together with the appropriate references. The listing is intended as a reference table for those readers that are interested in further reading on some specific problem area. The listing is divided into two classes of problems, i.e. *basic problems* and *engineering problems*. This division is rather arbitrary. In particular for the placement problem the border between the two classes is very vague. However, it serves our purpose here. The listing is a sample from the existing literature and by no means pretends to be complete. In particular in the field of VLSI design and physical engineering, there are many papers that are not mentioned here.

5.2.1 Basic Problems

Travelling Salesman Problems
Aarts & Van Laarhoven [1988], Aarts, Korst & Van Laarhoven [1988], Bonomi & Lutton [1984], Černy [1985], Felten, Karlin & Otto [1985], Golden & Skiscim [1986], Johnson, Aragon, McGeoch & Schevon [1988], Kirkpatrick [1984], Kirkpatrick & Toulouse [1985], Van Laarhoven [1988], Randelman & Grest [1986], Rossier, Troyon & Liebling [1986], Skiscim & Golden [1983], Sourlas [1986].

Graph Partitioning Problems
Aarts & Van Laarhoven [1985b], Johnson, Aragon, McGeoch & Schevon [1987], Kirkpatrick [1984], Van Laarhoven [1988], Sheild [1987].

Matching Problems
Lutton & Bonomi [1986], Sasaki & Hajek [1988], Telley, Liebling & Mocellin [1987], Weber & Liebling [1986].

Quadratic Assignment Problems
Bonomi & Lutton [1986], Burkard & Rendl [1984], Wilhelm & Ward [1987].

Linear Arrangement Problems
Bhasker & Sahni [1987], Nahar, Sahni & Shragowitz [1985; 1986].

Graph Colouring Problems
Chams, Hertz & De Werra [1987], Johnson, Aragon, McGeoch & Schevon [1988], Morgenstern & Shapiro [1986].

Scheduling Problems
Abramson [1987], Chen, Wong & Ping [1987], Eglese & Rand [1987], Van Laarhoven, Aarts & Lenstra [1988], Perusch [1987].

5.2.2   Engineering Problems

VLSI Design

- Placement Problems
Banerjee & Jones [1986], De Bont, Aarts, Meehan & O'Brien [1988], Casotto & Sangiovanni-Vincentelli [1987], Casotto, Romeo & Sangiovanni-Vincentelli [1987], Darema-Rogers, Kirkpatrick & Norton [1987], Devadas & Newton [1986], Grover [1986], Jepsen & Gelatt [1983], Kravitz & Rutenbar [1987], Otten & Van Ginneken [1984], Rose, Blythe, Snelgrove & Vranesic [1986], Sechen & Sangiovanni-Vincentelli [1985], Siarry, Bergonzi & Dreyfus [1987], Storer, Becker & Nicas [1985], Wong & Liu [1986].

- Routing Problems
Leong [1986], Leong & Liu [1985], Leong, Wong & Liu [1985], Vecchi & Kirkpatrick [1983].

- Array Logic Minimization Problems
Devadas & Newton [1986], Fleisher, Giraldi, Martin, Phoenix & Tavel [1985], Gonsalves [1986], Lam & Delosme [1986], Leong [1986], Rowen & Hennessy [1985], Wong, Leong & Liu [1986].

- Testing Problems
Distante & Piuri [1986], Ligthart, Aarts & Beenker [1986].

Facilities Layout
Sharpe & Marksjo [1985], Sharpe, Marksjo, Mitchell & Crawford [1985].

Image Processing
Carnevalli, Coletti & Patarnello [1985], Geman [1987], Geman & Geman [1984; 1986], Murray & Buxton [1987], Paxman, Smith & Barrett [1984], Ripley [1986], Smith, Barrett & Paxman [1983], Sontag & Sussmann [1986], Wolberg & Pavlidis [1985].

Code Design
Beenker, Claasen & Hermens [1985], Bernasconi [1987], El Gamal, Hemachandra, Shperling & Wei [1987], Van Laarhoven, Aarts, Van Lint & Wille [1988], Wille [1987].

Biology
Anderson [1983], Dress & Kruger [1987], Goldstein & Waterman [1987], Lundy [1985].

Physics
Lyberatos, Wohlfarth & Chantrell [1985], Nicholson, Chowdhary & Schwartz

[1984], Rothman [1985], Semenovskaya, Khachaturyan & Khachaturyan [1985], Telley, Liebling & Mocellin [1987], Wooten, Winer & Weaire [1985], Wille [1986b], Wille & Vennik [1985].

Finally, we mention that the simulated annealing algorithm has also been applied to optimization problems with continuous variables; for instance see the work of Bohachevski, Johnson & Stein [1986], Corana, Marchesi, Martini & Ridella [1987], Geman & Hwang [1986], Khachaturyan [1986], and Vanderbilt & Louie [1984]. In this context, the relation with the physical process of *Brownian motion* and the *Langevin equation* is of special interest [Geman & Hwang, 1986; Gidas, 1985b].

5.3   General Performance Experiences

The succes of an approximation algorithm depends on a number of aspects. We mention

- performance, i.e. the running time and error; see also Section 4.3,

- ease of implementation, and

- applicability and flexibility.

In the case of the simulated annealing algorithm we make the following remarks on the last two items. It is apparent that the simulated annealing algorithm is very simple and easy to implement. Implementation of the algorithm, for instance for the problems discussed in Section 5.1, typically takes only a few hundred lines of computer code. Our experience shows that implementations for new problems often take only a few days and in most cases existing programs, written for another problem, can be efficiently used.

With respect to applicability and flexibility we refer to the extensive list of Section 5.2, indicating that the algorithm is indeed applicable to a wide variety of problems and that it can handle variations of a problem. However, one must bear in mind that it is not always trivial to apply the algorithm to a given problem. It is sometimes necessary to reformulate the problem or transform it into an equivalent or similar problem, before the simulated annealing algorithm can be applied successfully, for instance in the case of the graph colouring problem; see Section 5.1.4.

With respect to performance, one traditionally distinguishes between the error and the running time of the algorithm; see also Section 4.3. In analysing the performance of an approximation algorithm one may distinguish between the following items.

- A theoretical performance analysis: deriving analytical expressions for the error and the running time. This can be done either for the worst-case or the average-case situation.

- An empirical performance analysis: deriving statistical ensemble averages on the error and the running time by running the algorithm a number of times for different initial solutions or for different problem instances. An empirical performance analysis refers to an average-case situation. An empirical worst-case performance analysis makes little sense since there is never any guarantee of the empirically obtained bound.

The literature presents the following theoretical worst-case performance results.

- Some cooling schedules can be executed with a running time bounded by a polynomial in the problem size; see Section 4.2.

- Upper bounds are known for the distances between the probability distribution of the solutions after a finite number of transitions, and the stationary distribution or the uniform distribution over the set of optimal solutions; see Section 3.5.

- For the maximum matching problem, Sasaki & Hajek [1988] give an upper bound on the error for a given set of problem instances.

Theoretical average-case performance results are not known from the literature except for the expected running time, derived by Sasaki & Hajek [1988] for the above mentioned matching problem. Results of a theoretical average-case analysis of the simulated annealing algorithm would be of great practical use, since they allow the practitioner to estimate the expected performance of the algorithm for a given problem instance. Therefore, we consider such an analysis as one of the most important open problems with respect to the simulated annealing algorithm. However, experience shows that it is also one of the hardest problems in this field [Van Laarhoven, 1988].

Finally, we mention a special class of theoretical analyses concentrating on the probabilistic description of the optimal cost for large randomly generated problem instances. Such *probabilistic value analyses* have been carried out with the simulated annealing algorithm for the travelling salesman problem [Bonomi & Lutton, 1984], the perfect minimal matching problem [Lutton & Bonomi, 1986], and the quadratic assignment problem [Bonomi & Lutton, 1986].

Empirical performance analyses of the simulated annealing algorithm have been the subject of many studies, e.g. by Aarts & Van Laarhoven [1988], Aarts, Korst & Van Laarhoven [1988], Golden & Skiscim [1986], Johnson, Aragon, McGeoch & Schevon [1987; 1988], Kirkpatrick [1984], Van Laarhoven [1988], Nahar, Sahni & Shragowitz [1986]; see also Sections 4.3 and 5.2. Despite the fact that the number of studies presented in the literature is large, it is still difficult to judge the simulated annealing algorithm on its true merits. This is predominantly due to the fact that many of these studies lack the depth required to draw reliable conclusions; for example, results are often limited to one single run of the algorithm, instead of taking the average over a number of runs; the applied cooling schedules are often too simple, and do not get the best out of the algorithm; results are often not compared to the results obtained with other (tailored) algorithms.

Lining up the most important and consistent results from the literature allows the following general observations.

- **Basic Problems** (predominantly based on Johnson, Aragon, McGeoch & Schevon [1987; 1988], and Van Laarhoven [1988])

  - For graph partitioning problems, the simulated annealing algorithm generally performs better, both with respect to error and running time, than the classical edge-interchange algorithms introduced by Kernighan & Lin [1970].

  - For a large class of basic problems, including the graph colouring, linear arrangement, matching, quadratic assignment, and scheduling problems, the simulated annealing algorithm finds solutions with an error comparable to the error of tailored approximation algorithms but at the cost of much larger running times.

  - For some basic problems such as the number partitioning problem and the travelling salesman problem, the simulated annealing algorithm is outperformed by tailored heuristics, both with respect to error and running time.

- **Engineering Problems** (folklore)

  For many engineering problems, for example problems in the field of image processing, VLSI design and code design, no tailored approximation algorithms exist. For these problems simulated annealing seems to be a panacea: it finds high-quality solutions, sometimes even better than the existing ones, e.g. for the football pool problem [Van Laarhoven, Aarts, Van Lint & Wille, 1988; Wille, 1987], or those obtained by a time-consuming manual process, e.g. for the placement problem [Sechen & Sangiovanni-Vincentelli, 1985]. For some problems however the running time can be very large. Sechen & Sangiovanni-Vincentelli [1986] for instance report CPU times of more than 84 hours for some instances of the placement problem.

- Comparing simulated annealing to time-equivalent local search on the same neighbourhood structure, i.e. repeating a local search algorithm with different initial solutions for an equally long time as the simulated annealing algorithm and keeping the best solution, reveals that the simulated annealing algorithm performs substantially better (smaller error). This difference becomes even more pronounced for larger problem instances [Aarts & Van Laarhoven, 1988; Van Laarhoven, 1988].

- Experience shows that the performance of the simulated annealing algorithm depends as much on the skill and effort that is applied to the implementation as on the algorithm itself. For instance the choice of an appropriate neighbourhood structure, of an efficient cooling schedule, and of sophisticated data structures allowing fast manipulations can substantially reduce the error as well as the running time. Thus, in view of this and considering the simple nature of annealing, there lies a challenge in constructing efficient and effective implementations of the simulated annealing algorithm.

CHAPTER 6

Parallel Simulated Annealing Algorithms

In the previous chapters we have studied the simulated annealing algorithm in great detail. Both theoretical and practical aspects have been discussed and the main conclusions that where reached can be formulated as follows. Advantages of the algorithm are its potential to find near-optimal solutions, its general applicability, its flexibility, and its ease of implementation. A disadvantage is the potentially burdensome amount of time required to converge to a near-optimal solution.

The amount of computational effort required by the simulated annealing algorithm strongly depends on the nature and the size of the optimization problem at hand. It ranges from a few seconds, e.g. for small instances of the travelling salesman problem, up to a few days, e.g. for large instances of the placement problem [Sechen & Sangiovanni-Vincentelli, 1985]. However, as problems inevitably increase in size, the situation with respect to the computational effort will worsen. Therefore, it is worthwhile to investigate possibilities of speeding up the algorithm, in order to keep computation times within reasonable limits. In this respect, the increasing availability of parallel machines offers an interesting opportunity to explore the possibilities of speeding up the simulated annealing algorithm. This is the main subject of this chapter.

6.1 Speeding up the Simulated Annealing Algorithm

Before we discuss a number of parallel versions of the simulated annealing algorithm, we briefly discuss a number of alternative approaches to the problem of speeding up the algorithm. An increase in speed may be achieved by means of the following three approaches:

(*i*) design of fast sequential algorithms,

95

(*ii*) hardware acceleration, and

(*iii*) design of parallel algorithms.

Below, we elaborate on these items in more detail.

(*i*) Improvements on the generation mechanism and/or the cooling schedule may result in more efficient implementations of the simulated annealing algorithm without deterioration of the quality of the final solution obtained.

An interesting approach along the lines of an improved generation mechanism is reported by Szu & Hartley [1987]. The authors present an annealing algorithm for optimization of continuous-valued functions, using a generation mechanism given by a Cauchy distribution instead of the usually used normal distribution. They claim that their generation mechanism leads to an inverse linear cooling rate instead of an inverse logarithmical cooling rate as was found for the Gaussian distribution. Whether the approach of Szu & Hartley is of significance for combinatorial optimization is questionable however.

Another example of an improved generation mechanism is the *rejection less method* introduced by Green & Supowit [1986]. These authors propose to generate new solutions with a probability proportional to the effect of a transition on the cost function. In this way, a subsequent solution is directly chosen from the neighbourhood of a given solution; no rejection of solutions will occur. This method leads to shorter Markov chains in a number of problems. However, the efficient use of the method depends strongly on some additional conditions on the neighbourhood structure, which, unfortunately, cannot be met by many combinatorial optimization problems.

With respect to cooling schedules, there exists in the literature a rich variety of alternative approaches. A detailed overview is given by Van Laarhoven & Aarts [1987]. Most of the reported cooling schedules are accompanied by the claim that they are faster than other existing ones. Unfortunately, many of the numerical comparisons are carried out on small sets of problem instances and often lack statistics, which reduces the value of the reported conclusions.

As already mentioned in Chapter 4, we hold the opinion that cooling schedules in general cannot substantially improve the algorithm's efficiency. However, we do not rule out the possibility of improving the efficiency by using cooling schedules that are tailored to a given problem or set of problems. An example of such an approach is given by Catthoor, DeMan & Vanderwalle [1988], who propose a cooling schedule for problems in which clustering occurs.

Finally, we mention that substantial improvements on computation times can also be achieved by more efficient implementations of the simulated annealing algorithm. For instance, Johnson, Aragon, McGeoch & Schevon [1987] mention that replacing the exponent $\exp(-\Delta f/c)$ in the acceptance criterion of (5.1) by the function $1 - \Delta f/c$ may speed up the algorithm by approximately 30%. The authors report similar results from the use of a look-up table for the value of the aforementioned exponent.

(ii) Hardware acceleration can be achieved by using dedicated hardware for evaluating time-consuming parts in the simulated annealing algorithm. For example, Iosupovici, King & Breuer [1983] use dedicated hardware for evaluating incremental wire lengths in a placement problem.

Another approach to hardware acceleration is to rewrite time-consuming parts of the algorithm in *micro code* to be executed on a fast general purpose *micro engine* attached to a host computer. Such an approach is pursued by Spira & Hage [1985], who report speed-up factors for placement problems ranging up to 20.

(iii) The key issue in designing parallel simulated annealing algorithms is to distribute the execution of the various parts in the simulated annealing algorithm over a number of communicating parallel processors. This is a promising approach to the problem of speeding up the execution of the simulated annealing algorithm, but it is by no means a trivial task. This is due to the intrinsic sequential nature of the algorithm, i.e. transitions are to be carried out one after the other.

Research on the design and analysis of parallel simulated annealing algorithms has evolved very quickly over the past few years. Both generally applicable algorithms and algorithms tailored to a given problem have been proposed. In the remainder of this chapter we discuss the basic aspects related to the design and analysis of parallel simulated annealing algorithms, and we briefly review the various approaches given in the literature.

The remainder of this chapter is organized as follows. In Section 6.2 we briefly review the existing machine models for parallel computing; in Section 6.3 we discuss a number of strategies for designing parallel simulated annealing algorithms, based on the present status of the literature; in Section 6.4 we present three generally applicable parallel simulated annealing algorithms that exhibit markedly different aspects, and compare the algorithms for their performance on the basis of a numerical average-case study.

The reader will observe that the description of the parallel algorithms presented in this chapter lacks the rigour of the description of the sequential algorithm presented in the previous chapters. Here, we confine ourselves to a discussion of the basic issues that connect simulated annealing and parallel computing. The reason for this is that the design of parallel simulated annealing algorithms is still in its infancy and it is moving rapidly. So far, there is no common agreement within the 'annealing community' as to a general approach and therefore we feel that the presentation of an introductory background, surveying the important features, is most appropriate here.

6.2   Parallel-Machine Models

Over the years many practical and theoretical architectures have been proposed for parallel computing. Some of these parallel-machine models have resulted in the construction of actual computer systems. Others serve as theoretical models for designing and analysing parallel algorithms, while their realization is not (yet) feasible.

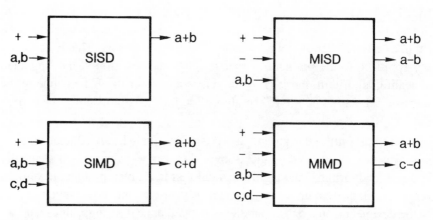

Figure 6.1: A classification of parallel-machine models according to Flynn [1966].

The most widely known classification of parallel-machine models is given by Flynn [1966]. Flynn distinguishes between the following four general classes; see also Figure 6.1:

- SISD (*Single Instruction, Single Data*): one instruction at a time is executed on one set of data. This class includes the classical sequential computers.

- SIMD (*Single Instruction, Multiple Data*): one instruction at a time is executed on multiple data sets. This class includes vector computers and array processors.

- MISD (*Multiple Instructions, Single Data*): multiple instructions at a time are executed on one set of data. This class has received little attention so far.

- MIMD (*Multiple Instructions, Multiple Data*): multiple instructions at a time are executed on multiple data sets. The processors in an MIMD machine are either synchronized, performing each successive set of instructions simultaneously, or unsynchronized, performing all instructions independently. Presently, this class receives much attention and many machines are being developed and built on the basis of this model.

In addition to architectural models, communicational models can be used to classify parallel-machine models. Here we follow Schwartz [1980], who distinguishes between the following two models:

- *Para computers*: processors have simultaneous access to a shared memory, allowing communication between two arbitrary processors in constant time. These machines are of great theoretical interest but their realization is not (yet) possible due to technological limitations.

- *Ultra computers*: processors communicate through a fixed interconnection network. Such a network can be represented by a graph, whose nodes correspond to processors and the edges to interconnections. Important quantities of such a graph are the maximum degree of the nodes and the diameter. To ensure fast communication through the network, the maximum degree should be bounded by a constant and the diameter should grow at most logarithmically with the number of nodes in the graph. Over the years a number of networks have been proposed. We mention the *two-dimensional mesh-connected network*, the *hypercube-connected network*, the *perfect-shuffle network* and the *binary-trees network*. All these networks have a small maximum degree and diameter [Schwartz, 1980].

The classification of parallel-machine models given above leads to the following four classes of parallel algorithms [Van Leeuwen, 1985].

- *Vectorized algorithms*: operating uniformly on vectors of data sets; the corresponding machines are SIMD machines: vector computers and array processors.

- *Systolic algorithms*: operating rhythmically on streams of data sets; corresponding machines: SIMD and synchronous MIMD machines.

- *Parallel processing algorithms*: operating on a set of parallel processors that communicate synchronously; corresponding machines: ultra computers and synchronous MIMD machines.

- *Distributed processing algorithms*: operating on a set of parallel processors that communicate asynchronously; corresponding machines: asynchronous MIMD machines, para computers, data-flow computers and neural networks.

The quality of a parallel algorithm is determined by a number of quantities, the most important one being the speed-up. The *speed-up* is defined as the running time of the sequential algorithm executed on one processor, divided by the running time of the parallel algorithm executed on a number of processors. The *processor utilization* or *efficiency* is defined as the speed-up divided by the number of processors used to execute the algorithm. Clearly, the best one can do with a parallel algorithm is to attain a speed-up equal to the number of processors and an efficiency equal to one. If the speed-up increases linearly with increasing number of processors, we say that the speed-up is linear. A linear speed-up implies that there is no saturation with an increasing number of processors.

6.3   Designing Parallel Annealing Algorithms

From Section 5.1 we recall that execution of the simulated annealing algorithm requires specification of the following quantities:

- a data structure representing the solution space,

- a cost function whose differences can preferably be calculated incrementally,

- a neighbourhood structure representing for each solution the set of (feasible) solutions that can be reached in one single step, and

- a cooling schedule.

The most time-consuming part in the execution of the algorithm is given by the generation of the sequence of trials that constitute a Markov chain; see Chapter 4. Hence, strategies for designing efficient parallel simulated annealing algorithms should focus on this part of the algorithm. Evaluation of a trial consists of the following four tasks.

1. Selecting a new solution from the neighbours of the current solution.

2. Calculating the difference in cost between these two solutions.

3. Deciding whether or not the new solution is to be accepted.

4. Replacing the current solution by the new solution if the new solution is accepted.

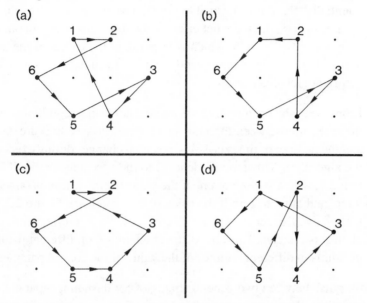

Figure 6.2: Four solutions of a 6-city travelling salesman problem. (a) Initial solution; (b) solution after applying an $N_2(4, 6)$ change to (a); (c) solution after applying an $N_2(5, 1)$ change to (a); (d) solution after applying an $N_2(5, 1)$ change to (b).

Closer examination of these tasks leads to the following two observations.

A. The first three tasks are essentially independent and can be executed in parallel for different trials without effecting the sequential nature of the algorithm.

   Executing the fourth task in parallel may create several problems. It not only may be impossible to replace accepted solutions in parallel - which basically would require a para computer - it also may lead to *erroneously calculated transitions*. Consider, for example, the solution of the 6-city instance of the travelling salesman problem shown in Figure 6.2a. The length of the tour equals $f = 1+2\sqrt{2}+3\sqrt{5}$. Now suppose one processor attempts the 2-change $N_2(4, 6)$, and another processor the

2-change $N_2(5, 1)$; for the definition of a 2-change see Example 1.3. The resulting tours are shown in Figures 6.2b and 6.2c, respectively. The corresponding tour lengths are $3 + 3\sqrt{2} + \sqrt{5}$, and $2 + 2\sqrt{2} + 2\sqrt{5}$, respectively, and are smaller than the length of the tour of Figure 6.2a. Hence, both processors will accept the proposed transitions through a local decision. However, implementation of two or more 2-changes on the same solution cannot be done simultaneously; it must be done sequentially. So, if it is decided to implement the $N_2(4, 6)$ first, followed by the $N_2(5, 1)$, then the net effect is the tour shown Figure 6.2d, with length $2 + 2\sqrt{2} + 3\sqrt{5}$, which is larger than the length of the original tour in Figure 6.2a. Consequently, correct local decisions may lead to incorrect global decisions.

B. From the typical nature of the simulated annealing algorithm we know that the ratio between the number of times tasks 1 to 3 are executed and the number of times task 4 is executed changes during execution of the algorithm. This ratio is identical to the acceptance ratio of Definition 2.7, which is close to one if the value of the control parameter c is large, and close to zero if the value of c is small; see Figure 2.2. Thus, the relative frequency of task 4 decreases with decreasing values of the control parameter. This allows the efficient use of different parallel algorithms in different regimes of the value of the control parameter.

We distinguish between two general strategies for designing a parallel simulated annealing algorithm, i.e. single-trial parallelism and multiple-trial parallelism.

In *single-trial parallelism*, the work involved in evaluating a single trial is divided over a number of processors (*function division*). The speed-up that can be achieved with this strategy strongly depends on the problem at hand. For instance, in the placement problem generating a new placement and calculating the corresponding difference in cost involves many operations that can be straightforwardly divided into a number of subtasks which then can be assigned to different processors [Casotto, Romeo & Sangiovanni-Vincentelli, 1987; Kravitz & Rutenbar, 1987]. For the travelling salesman problem however it is very hard to find such a division. Unfortunately, this is the case for a number of combinatorial optimization problems and therefore application of single-trial parallelism is restrictive.

In *multiple-trial parallelism*, trials are evaluated in parallel. A straightforward approach is given by the following algorithm: processors simultaneously generate their own transitions from the same current solution. This process is repeated until one of the processors accepts a new solution. Then all processors are halted and reset; the current solution is replaced by the newly accepted solution and this solution is communicated to all processors. After this, the processors continue generating new solutions from the new current solution. If there are more processors at the same time that have accepted their new solution, then one of those solutions is selected by an arbiter to become the new current solution, and the other new solutions are discarded. With this approach, the speed-up is small (≈ 1) at large values of the control parameter, since each new solution will be accepted allowing only one processor at a time to be active. However, as the value of the control parameter decreases, the speed-up will increase and eventually become proportional to the number of processors if no new solutions are accepted anymore.

Most of the work on the design of parallel simulated annealing algorithms presented in the literature is based on multiple-trial parallelism. We distinguish between two classes: *general algorithms* and *tailored algorithms*, where the adjectives refer to 'generally applicable' and 'tailored to a given problem', respectively.

Only a few general algorithms are known from the literature. Two examples are presented by Aarts, De Bont, Habers & Van Laarhoven [1986]. The first algorithm, known as the *systolic algorithm*, generates Markov chains in parallel, i.e. each of the available processors generates its own Markov chain, and during this generation, information is transferred from a processor generating a given Markov chain, to the processor generating the succeeding Markov chain. The second algorithm, known as the *clustering algorithm*, uses all available processors to generate a single Markov chain. General algorithms are readdressed in Section 6.4.

Most parallel simulated annealing algorithms presented in the literature are tailored algorithms. In many of these algorithms parallelism is achieved by partitioning the data structures, representing the solution space as it typically exists for the problem at hand. For example, Felten, Karlin & Otto [1985] show that a hypercube-connected network of processors can efficiently solve travelling salesman problems by simulated annealing. In their approach, the cities in the travelling salesman problem are evenly distributed over the available processors. Two actions are combined: each processor determines a (sub)tour by simulated annealing and in between these calculations cities are exchanged between the processors along the edges of the hypercube. Efficiencies of up to 1 are reported.

Other examples of parallel annealing algorithms applying partitioning approaches to placement and routing problems in VLSI design are reported by Darema-Rogers, Kirkpatrick & Norton [1987], Banerjee & Jones [1986], Casotto, Romeo & Sangiovanni-Vincentelli [1986], and Devadas & Newton [1986]. These problems are suitable for partitioning, because they are defined on one- or two-dimensional maps that allow spatial division. Similarly to the approach followed by Felten, Karlin & Otto [1985], the partitioning used in these algorithms is dynamic, i.e. the partitioning of the data structure over the available processors varies in time.

As mentioned earlier (observation B), the simulated annealing algorithm exhibits markedly different characteristics in different regimes of the value of the control parameter. A number of authors use these differences in constructing adaptive strategies, e.g. Aarts, De Bont, Habers & Van Laarhoven [1986], De Bont, Aarts, Meehan & O'Brien [1988], and Kravitz & Rutenbar [1987]; see also Section 6.4.2. These strategies are based on the use of different algorithms in the different regimes. As an example we discuss the approach pursued by Kravitz & Rutenbar [1987]. Their approach is tailored to placement problems and combines the following two algorithms.

In the large-value regime they use the *single-move decomposition algorithm*, which is based on single-trial parallelism. It divides the tasks involved in computing a single trial into several subtasks that can be evaluated in parallel. As we already mentioned, this is a sensible approach to placement problems, since evaluation of transitions takes substantial computational efforts.

In the small-value regime they use the parallel-moves algorithm which is based on multiple-trial parallelism. This algorithm uses the concept of *serializable subsets* of transitions, i.e. sets of transitions that do not interact. Determination of serializable subsets for a given problem instance is difficult and needs to be repeated often. Therefore, they adopt the idea that serializable subsets can be obtained by rejecting accepted transitions, i.e. only the first accepted transition found during the parallel evaluation of transitions is counted; all the other accepted transitions are aborted. Thus, if N transitions are calculated out of which M are rejected, then $N - M + 1$ transitions are counted as trials in the generation of a Markov chain.

The speed-up of the single-move decomposition algorithm is found to be constant (≈ 2) as a function of the control parameter. The speed-up of the parallel-moves algorithm is small (≈ 1) in the large-value regime and large (≈ 3.5) in the small-value regime of the control parameter; the numbers refer to an implementation on a 4-processor computer system. Thus, at some value of the control parameter it becomes favourable to switch from the single-move decomposition algorithm to the parallel-moves algorithm. The switching value is determined during execution of the algorithm using a probabilistic model. We mention that the parallel-moves algorithm is a general algorithm in nature, since it is not tailored to a given problem.

Most of the parallel annealing algorithms we have been discussing so far, strictly follow the sequential simulated annealing algorithm in the sense that they only execute those tasks in parallel that are independent, i.e. tasks 1-3; see also observation A. These algorithms are suited for execution on synchronous MIMD machines and belong to the class of parallel processing algorithms; see Section 6.2. Furthermore, it is important to note that, as a consequence of the fact that the sequential nature of the algorithm remains basically unaffected, the asymptotic convergence properties discussed in Chapter 3 also hold for this class of parallel algorithms.

Another class of algorithms is based on multiple-trial parallelism, employing replacements of solutions that are accepted in parallel. Hence, these algorithms allow for erroneously calculated transitions; see observation A. Hereafter, these algorithms are called *error algorithms*. For example, Casotto, Romeo & Sangiovanni-Vincentelli [1987] achieve parallelism by partitioning the set of cells in a macro-cell placement problem into as many subsets as there are processors available, and then assigning each subset to a different processor. The processors are allowed to run asynchronously, as they independently calculate transitions among placements, assuming that all cells assigned to other processors are fixed, thereby allowing erroneously calculated transitions. Similar approaches are reported by Rose, Blythe, Snelgrove & Vranesic [1986], and Darema-Rogers, Kirkpatrick & Norton [1987]. It is argued by the authors that the errors, introduced by the erroneously calculated transitions, disappear as the value of the control parameter becomes small. This can be understood from the following intuitive argument. If the value of the control parameter becomes small, the relative frequency of accepted transitions, and thus of simultaneous replacements, becomes small. Hence, erroneously accepted transitions may be corrected at small values of the control parameter. As we will see in Section 6.4.3, an additional condition for convergence is given by the requirement that transitions be local, i.e. the difference between a current and a new solution should be confined to a small local rearrangement.

From their numerical experiments, Casotto, Romeo & Sangiovanni-Vincentelli [1987] conclude that the global performance of the parallel annealing algorithm, applied to placement problems, is not affected by introducing erroneously calculated transitions. Darema-Rogers, Kirkpatrick & Norton [1987] and Rose, Blythe, Snelgrove & Vranesic [1986] report similar observations but under the additional condition that the correct value of the cost function is recalculated at the end of each individual Markov chain.

An important feature of error algorithms is that trials can be executed asynchronously, allowing execution on asynchronous MIMD machines. Thus, error algorithms can be viewed as distributed processing algorithms. Furthermore, we mention that, due to the erroneously calculated transitions, asymptotic convergence of error algorithms can no longer be proved along the lines presented in Chapter 3. So far, we are not aware of any rigorous convergence proof for this class of algorithms. Numerical evaluations merely indicate that, for some problems, error algorithms converge to near-optimal solutions. However, for other problems, no convergence is attained at all. We readdress this subject in Section 6.4.3.

Most of the algorithms presented in the literature and discussed in this section have been implemented on small-scale parallel MIMD machines with up to 16 processors. Efficiencies up to 0.8 are reported. Typical for these algorithms is that their implementation is limited to small-scale parallel machines. Implementations on large-scale parallel machines are either infeasible or lead to a fast decrease of efficiency.

An example of an implementation of a parallel simulated annealing algorithm on a large-scale parallel machine is presented by Casotto & Sangiovanni-Vincentelli [1987]. Their algorithm is based on multiple-trial parallelism and partitioning techniques tailored to placement problems, and is implemented on a *connection machine* [Hillis, 1985] using 16,384 processors. The speed-up obtained with this approach is however small. Compared to a VAX-11/780 computer, their present implementations do not achieve any speed-up at all, but they anticipate that a more sophisticated implementation would yield a speed-up equal to 90.

6.4 General Algorithms

In the previous section we discussed a number of parallel simulated annealing algorithms known from the literature. Most of these algorithms are tailored to a given problem, using functional-division or spatial-division techniques. The extent to which these algorithms can be successfully applied to other problems strongly depends on the possibilities of finding suitable partitions. As was pointed out in the previous section, some problems are very suitable for this approach, e.g. VLSI placement and routing problems, whereas others are less well suitable, e.g. the travelling salesman problem. This creates a need for more general algorithms that can be successfully applied to a wide range of optimization problems.

In this section we discuss three examples of more general algorithms, i.e. the *division algorithm*, the *clustering algorithm* and the *error algorithm*. The algorithms are conceptually simple, easy to implement, and they all use multiple-trial parallelism.

6.4.1 The Division Algorithm

In the division algorithm, parallelism is obtained by dividing the effort of generating a Markov chain over the available processors. Let p denote the number of processors and L the length of a Markov chain. Then a Markov chain is divided into p subchains of length $\lfloor L/p \rfloor$, and each processor generates its own subchain. One may now pursue two different strategies. In the first strategy there is no communication in between the generation of consecutive Markov chains, i.e. each processor continues the generation of a subsequent subchain with the solution given by the outcome of the last trial of the preceding subchain, obtained by the same processor; see Figure 6.3a. In the second strategy the best solution found by the processors that each generate their own subchain is used as the 'outcome' of the Markov chain constituted by the p subchains. In between subsequent Markov chains this solution is communicated to all processors and used as initial solution for generating the subchains constituting the next Markov chain; see Figure 6.3b. The first strat-

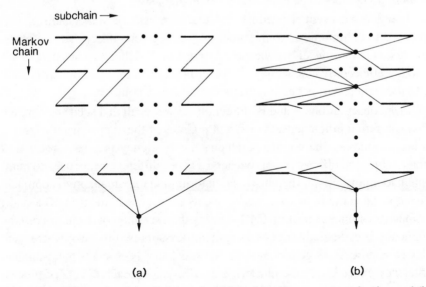

(a)   (b)

Figure 6.3: Schematic of the division algorithm, (a) without communication and (b) with communication.

egy returns p solutions - one for each processor - the best of which is chosen as the final solution. The second strategy returns one solution. The initial solution for both strategies may be chosen to be either the same or different for all subchains.

Compared to the sequential algorithm, using a similar cooling schedule, one may expect for the division algorithm a linear speed-up and an efficiency close to 1. In Section 6.4.4 we will put this to a numerical test.

As a prelude to the discussion of the results obtained with this test, we mention that the division algorithm performs equally well for both strategies mentioned above, i.e. no significant differences, both in the error and the running time, were observed from a number of runs on different instances of the placement and travelling salesman problem. The advantage of the first strategy is however the absence of communication, which allows the algorithm to be executed on asynchronous MIMD machines.

6.4.2   The Clustering Algorithm

In the clustering algorithm parallelism is achieved by letting all available processors cooperatively generate the same Markov chain; see also Aarts, De Bont, Habers & Van Laarhoven [1986]. This can be done either with or without division into subchains. 'Clustering' refers to the action of combining two or more processors to generate trials simultaneously.

In order to prevent erroneously calculated transitions, a mechanism is introduced to communicate to the processors in a given cluster that one of the processors in the cluster has accepted a transition and that the current solution is to be replaced by a new one. The other processors then abort their present activities and restart generating trials from the new solution.

From observations A and B of Section 6.3 we recall that such an approach yields different efficiencies in different regimes of the control parameter, i.e. at large values of the control parameter the efficiency is small, whereas at small values the efficiency approaches 1. These differences are used to combine two algorithms in the clustering algorithm using an adaptive approach. This is done in the following way. The clustering algorithm starts off as a division algorithm; see Section 6.4.1. During the execution of the algorithm the efficiency is estimated for clustering the processors two-by-two, where each cluster of processors generates the same subchain. As soon as the estimated efficiency exceeds 0.5, the clustering is effectuated. Each cluster of processors then generates a subchain which is twice as long as the subchain before clustering. The clustering process is continued, i.e. each time the estimated efficiency for a larger cluster - four-by-four, eight-by-eight, etc. - exceeds the value of 0.5, further clustering is effectuated until eventually all processors simultaneously generate a Markov chain of full length. The efficiency of clustering can be easily estimated by using the acceptance ratio and concepts from queueing theory [Aarts, De Bont, Habers & Van Laarhoven, 1986]. On the basis of these estimates it was derived that the clustering algorithm, similarly to the division algorithm discussed in Section 6.4.1, yields an efficiency equal to 1. We come back to this in Section 6.4.4.

Neither the division nor the clustering algorithm do affect the basic convergence properties of the sequential simulated algorithm described in Chapter 3. As mentioned before, these algorithms closely follow the sequential algorithm and therefore it can be easily verified that they converge asymptotically to the set of optimal solutions. Furthermore, it was derived by Aarts, De Bont, Habers & Van Laarhoven [1986] that the clustering algorithm has a polynomial-time complexity if a similar cooling schedule is applied as the one presented in Section 4.2. Moreover, from numerical experiments, the authors show that the error is more or less constant as a function of the number of processors; see also Section 6.4.4. Similar arguments can be used to conjecture the same behaviour for the division algorithm.

6.4.3 The Error Algorithm

In the error algorithm parallelism is achieved by letting all available processors generate the same Markov chain. No division into subchains and no communication is used. Therefore, the error algorithm is suitable for execution on an asynchronous MIMD machine.

The notion 'error' refers to the fact that the algorithm allows erroneously calculated transitions. Basically, all actions needed to evaluate a trial are carried out in parallel, also replacements. However, parallel execution of replacements would require a para computer, but such machines are not (yet) available; see Section 6.2. Hence, implementation on an asynchronous ultra computer using a global memory requires sequential execution of replacements. This requires the use of an *arbiter* which determines the sequence in which simultaneously handed-in replacements are evaluated. This however is only a minor modification which has no effect on the performance of the algorithm.

It should be noted that the clustering algorithm described in Section 6.4.2 becomes identical to the error algorithm by running the former algorithm with exactly one cluster, containing all processors, and the communication between processors being switched off.

As a result of the presence of erroneously calculated transitions, the error algorithm no longer follows the sequential simulated annealing algorithm strictly, and consequently the asymptotic convergence properties of the sequential algorithm derived in Chapter 3 can no longer be assumed to hold.

So far, only empirical results are available on convergence properties of the error (or related) algorithms; see the results obtained by Casotto, Romeo & Sangiovanni-Vincentelli [1987], Darema-Rogers, Kirkpatrick & Norton [1987], and Rose, Blythe, Snelgrove & Vranesic [1986]; see also Section 6.4.4. A mathematical analysis of the asymptotic convergence properties of error algorithms and a formulation of the conditions for convergence, either necessary or sufficient, are not known from the literature. They would however be of great use for understanding and analysing these algorithms.

Clearly, the advantage of the error algorithm is the absence of communication requirements, enabling simple and straightforward implementation of the algorithm on asynchronous MIMD machines.

As a prelude to the conclusions obtained in Section 6.4.4 from numerical experiments with the error algorithm for the travelling salesman problem, we can already state that the efficiency of the error algorithm, when applied to the travelling salesman problem, is small. Furthermore, it was observed for the travelling salesman problem that with larger numbers of processors no convergence could be attained at all! This is not in agreement with the conclusions obtained by Casotto, Romeo & Sangiovanni-Vincentelli [1987], Darema-Rogers, Kirkpatrick & Norton [1987], and Rose, Blythe, Snelgrove & Vranesic [1986]. These authors report large efficiencies, when applying error algorithms to placement problems.

This discrepancy can be explained from the type of problems the algorithm was applied to, using the following intuitive arguments. The generation mechanisms used for solving placement problems with simulated annealing induce only local rearrangements; see for instance Section 5.1.5. This reduces the probability that two simultaneously accepted transitions are indeed erroneously calculated. In the case of a 2-change generation mechanism, which was used for the travelling salesman problem, rearrangements on the average involve a substantial number of cities in a tour - the cities whose order has to be reversed - and are therefore by no means local. Consequently, the probability that two simultaneously generated transitions are erroneously calculated is considerable and clearly hampers convergence. From this we conclude that successful application of the error algorithm requires the use of generation mechanisms that induce only local rearrangements. This makes the algorithm less general than might have been expected.

6.4.4   Parallel Implementation and Numerical Results

In this section, we briefly discuss some results obtained from numerical experiments with the three parallel simulated annealing algorithms presented in Sections 6.4.1-6.4.3. The algorithms were applied to the EUR100 instance of the travelling salesman problem given in the Appendix, and were implemented on an experimental multi-processor system.

The system is a general-purpose MIMD machine consisting of 8 processors, a data bus, a graphics subsystem, a global memory, and an interface to a host computer. The processors each have an MC68020 micro processor, running at 16 MHz, with a floating-point co-processor and 1 Mbyte local memory. The local memory is dual-ported, i.e. it is possible to perform read and write operations on the memory of other processors. The data bus is 32 bits wide, integers and reals in one transfer, and the global memory consists of 8 Mbyte. The system is operated through a host computer. Information exchange between the processors takes place by means of a global memory. Mutual exclusion is achieved by using a semaphore s and the procedures, **SIGNAL**(s) and **WAIT**(s). An elementary action of the **SIGNAL**(s) and **WAIT**(s) procedures is an indivisible read-modify-write action of the TAS field of the semaphore for which the TAS instruction of the MC68020 is used. We mention that the average CPU time of a single iteration on the multi-processor system is about the same as on a VAX-11/780 computer.

For all three algorithms the conceptually simple cooling schedule of Section 4.1 was used. The initial value of the control parameter was calculated according to (4.5), using the parameter χ_0 as an input. The decrement rule was given by $c_{i+1} = \alpha c_i$ and the Markov chain length was chosen equal to a constant L for all Markov chains. The algorithm was terminated if the same value of the cost function was obtained for K subsequent Markov chains. In our implementations we used the following parameter values: $\chi_0 = 0.9$, $\alpha = 0.9$, $L = 5000$ and $K = 10$.

The performance of the parallel simulated annealing algorithms was analysed by empirically investigating the average-case behaviour of the error $\mathcal{E}$ and the speed-up s as a function of the number of processors p. The numerical data were obtained by running the various algorithms a number of times for the same set of parameters in the cooling schedule, using different initial solutions; see also Section 4.3.

Figures 6.4 and 6.5 show the average error and speed-up as a function

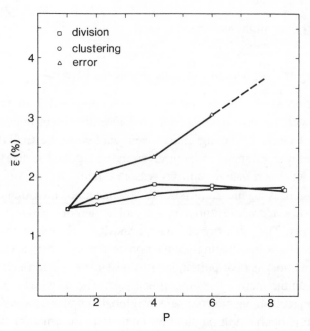

Figure 6.4: Average error as a function of the number of processors for the division, the clustering and the error algorithm (problem instance: EUR100; $\chi_0 = 0.9, \alpha = 0.9, L = 5000$ and $K = 10$).

of the number of processors for the three parallel simulated annealing algorithms. It is observed that both the division and clustering algorithms show hardly any increase of the average error as the number of processors increases, whereas the speed-up is about linear. Hence, for the small-scale parallel implementations of these algorithms, an efficiency close to 1 can be achieved without loss of effectivity. Estimates for the clustering algorithm however indicate that the efficiency decreases if larger numbers of processors are used. This is because of saturation phenomena in the communication network of the multi-processor systems we have been using.

With respect to the error algorithm we observe a rather poor performance: as the number of processors increases, the average error increases, while the efficiency decreases. For larger numbers of processors ($p \geq 8$) no convergence could be achieved. As mentioned before, this can be explained from the nature of the problem at hand; see Section 6.4.3.

In conclusion we can state that the division and the clustering algorithms perform well, but that the performance of the error algorithm is poor. As a final remark, we mention that the use of these algorithms is limited to imple-

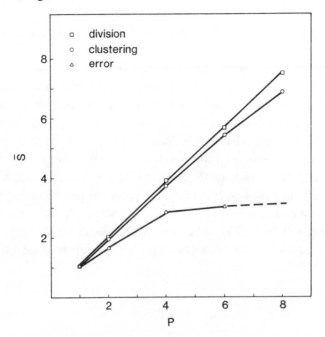

Figure 6.5: Average speed-up as a function of the number of processors for the division, the clustering and the error algorithm (problem instance: EUR100; $\chi_0 = 0.9$, $\alpha = 0.9$, $L = 5000$ and $K = 10$).

mentations on small-scale parallel machines. For the division and the clustering algorithms this is obvious, since the division of the Markov chains into subchains is limited as the subchains must be of reasonable length. The suitability of the error algorithm for implementations on multi-parallel systems is questionable. For placement problems the algorithm seems to perform well; for the traveling salesman problem the performance is however poor.

In this chapter, we have touched upon some aspects of designing and analysing parallel simulated annealing algorithms. Characteristic of the algorithms we have been discussing is the fact that they all, in one way or another, use the sequential simulated annealing algorithm as a starting point. Most parallel simulated annealing algorithms follow the sequential algorithm rather strictly, e.g. the division and clustering algorithms. But even in the presence of erroneously calculated transitions, as in the error algorithm, the deviation from the sequential algorithm is only minor. The major limiting factor of these algorithms is that they are not suited for execution on massively parallel machines. One exception may be the functional-division algorithm presented by Casotto & Sangiovanni-Vincentelli [1987], which runs on a connection machine with

16,384 processors. However, even in this case the reported efficiencies are very small.

Our general feeling is that the sequential nature of the simulated annealing algorithm in its present form is a drawback that greatly hampers the design of massively parallel simulated annealing algorithms. In our opinion, the design of such algorithms requires an approach that differs substantially from those presented in this chapter. In Part II of this book we discuss such an approach, which is based on the paradigm of neural computing; see Chapter 7. The resulting model of Boltzmann machines, presented in Chapter 8, allows for efficient massively parallel execution of the simulated annealing algorithm in a neural network, i.e. an asynchronous network of very simple computing elements, which can be viewed as a special-purpose asynchronous MIMD machine. The relation of this approach to combinatorial optimization is discussed in Chapter 9.

PART II

Boltzmann Machines

CHAPTER 7

Neural Computing

Recent advances in the design and manufacturing of integrated circuits have brought the construction of parallel computers, consisting of thousands of individual processing units, within our reach. A direct consequence of these technological advances is the growing interest in computational models that support in a natural way the exploitation of massive parallelism.

Until recently, massive parallelism did not receive much attention and we are just starting to understand its potential benefits. One of the few 'places' where massive parallelism is generally considered to be of dominant importance is the human brain. Thus, it seems natural to study computational models that are inspired by an analogy with the *neural network* of human brains. The resulting research field is called *neural computing*.

Presently, neural computing is attracting much attention from the scientific community, and considerable efforts are being made in exploring both theoretical and practical aspects of the many neural network models that have been proposed; cf. the special issues of *Computer* [1988] and the *Communications of the ACM* [1988], the first issue of *Neural Networks* [1988], and the *Proceedings of the International Conference on Neural Networks* [1987].

Neural computing can be viewed as the area in computer science, in particular artificial intelligence, that investigates computational models aiming at achieving human-like performance in computer systems via dense interconnection of very simple processing elements [Denker, 1986; Grossberg 1988a; Kohonen, 1988]. The topology and functioning of these networks is based on our present understanding of biological neural systems such as the human brain. The cortex of the human brain comprises a huge number ($\approx 10^{10}$) of cells, the neurons, connected in a complex way by synaptic junctions on the axons [Braitenberg, 1978]. Although the biochemical mechanism governing the metabolism of neurons is extremely intricate, and not yet fully understood, the information processing by a single neuron is straightforward. The information is viewed as being contained in trains of action potentials propagating along the axons. Furthermore, Braitenberg assumes that a global state of the cortex is fully determined by the individual states of the pyramidal neurons, which account for about $\frac{2}{3}$ of the population of neurons in the cortex. This view is adopted in most neural network models. It implies that a neural network can be thought of as a large assembly of identical elements, each characterized by an internal state that is determined by the activity of the elements it is connected to.

The greatest potential of neural computing is in the areas where high computation rates are required and present computer systems perform poorly, for instance speech and image processing [Denker, 1986; Grossberg, 1988a; Kohonen, 1988]. The basic paradigm of neural computing is the belief that massive parallelism is essential for high performance and fault tolerance.

The potential benefit of neural computing extends beyond purely technical advantages, however. Incorporating in a computer system human-like abilities such as associative capabilities and learning is considered to be of great importance for simplifying the communication between man and machine. Up to now, people always have had to adjust themselves considerably when communicating with computers. Making computers more intelligent may greatly enhance their user-friendliness.

Many models in neural computing have the capability of learning through adaptation of connection strengths between the processing units. In this way the performance of the system can be improved in time by using training data. The ability of learning is essential in areas such as speech and image processing. Moreover, learning and eventually relearning provide a kind of robustness since they compensate for minor variations and damage in connections and processing elements.

One of the recently introduced models in the field of neural computing is the Boltzmann machine [Hinton & Sejnowski, 1983; Hinton, Sejnowski & Ackley, 1984; Ackley, Hinton & Sejnowski, 1985]. The Boltzmann machine combines interesting properties from both neural computing and simulated annealing, resulting in a powerful computational model exploiting massive parallelism in a natural way.

The following chapters of this book are devoted to an extensive discussion of the Boltzmann machine. Both theoretical aspects, including convergence properties for different models of parallelism, as well as possible applications of the model in the fields of combinatorial optimization, pattern recognition and learning are treated in detail.

The remainder of this chapter aims at presenting the reader with a short introduction to the field of neural computing, by briefly discussing some general properties of neural network models and by giving a short historical overview.

7.1 Man versus Machine

Comparing the architectural structure of the human brain with that of a conventional computer reveals that, at a very low level, both structures may be considered as networks, consisting of conceptually simple nodes connected in some way: a brain is a network of neurons, a computer is a network of transistors. Furthermore, the comparison highlights a number of intriguing differences. Here, we mention two striking ones. Firstly, there is a large difference in connectivity. The nodes in a computer are usually only sparsely connected. A transistor is estimated to be directly connected, on the average, to less than 10 other transistors, while neurons in a human brain have up to 10,000 synaptic connections with other neurons [Braitenberg, 1978]. Secondly, there is a striking difference in response time or switching time of the nodes in the network. The time a neuron needs to respond to a stimulus is in the millisecond range. The switching time of a transistor in a present-day VLSI circuit is in the order of nanoseconds. So, roughly speaking, transistors operate a million times faster than neurons.

Despite the relative slowness of neurons, human beings are able to recognize images and understand speech in fractions of a second. Such classification and pattern recognition tasks, if they can be handled by computers at all, will be done so at the expense of large computation times [Devijver & Kittler, 1987; Young & Calvert, 1974]. It is commonly believed that the explanation of this paradox lies in the use of parallelism. A neural network is a highly parallel device, while a conventional computer, based on the classical Von Neumann model, is inherently sequential. In the latter type of machine, instructions are fetched from the main memory and executed by the central processing unit (CPU), essentially one at a time. Consequently, only a very small number of transistors is active at the same time, and efficient execution is greatly hampered by the communication between the CPU and the main memory. This drawback is generally known as the *Von Neumann bottleneck*.

In neural networks, memory is totally distributed over the network, allowing massively parallel execution. Moreover, as a result of the massive distribution of computation capabilities, the Von Neumann bottleneck is circumvented in these networks.

These intriguing differences, together with the ever increasing demands for fast computing, have, for quite some time, motivated people to investigate computational models and computer architectures that have a close resemblance with neural networks. Some people hope to be able to build faster computers in this way. Others try, by studying these models, to enlarge their understanding of the fundamental principles that underlie the functioning of neural networks and to give answers to questions as to how intelligent behaviour can emerge from such types of network models. From this, it is obvious that interest in these types of models extends over a number of disciplines, including computer science, artificial intelligence, cognitive science, and psychology, as well as application areas such as image and speech processing.

Neural computing aims at achieving human-like behaviour of computer systems by establishing an analogy with the human brain (analogy on a hardware level). The question whether such a hardware analogy is necessary to be able to achieve human-like behaviour is by no means settled. Much research within the field of artificial intelligence aims at achieving intelligent behaviour by means of an analogy with the human mind (analogy on a software level). By using specific formalisms one tries to simulate the reasoning process of human beings with moderate success, leading to systems that behave intelligently in a very restricted domain. For an overview see the work of Barr & Feigenbaum [1981; 1982] and Cohen & Feigenbaum [1982].

Many researchers, however, believe that the use of massive parallelism might very well be a necessary condition for building systems that show a more general intelligence. This opinion seems especially valid for the aforementioned areas such as pattern recognition, where computational requirements have always been a major limitation [Devijver & Kittler, 1987; Young & Calvert, 1974].

7.2 Connectionist Models

Neural computing aims at constructing models that follow the analogy with neural networks in the human brain. These models can be roughly divided into two classes, i.e. (i) models emphasizing computational aspects and (ii) models emphasizing biological fidelity. Here, we are predominantly interested in the first class of models, which we prefer to denote with the term *connectionist models*, instead of the more general term *neural network models*.

The term connectionist models was introduced by Feldman & Ballard [1982] and stresses the importance of performance through connections as opposed to computational speed of individual processors. Other names used to denote this class of models are *artificial neural net models*, *parallel distributed processing models* and *neuromorphic models*. Extensive overviews of the various models that have been developed during the recent years can be found in the excellent papers by Grossberg [1988b] and Hinton [1987], and in the books by Amari & Arbib [1982], Grossberg [1988a], and Rumelhart, McClelland & the PDP Research Group [1986].

For the many existing different connectionist models, a number of similarities can be identified. These similarities cover the basic aspects of the existing models and can be formulated as follows.

- All connectionist models assume that computations are carried out by a network of simple neuron-like processing elements, called units, that are densely interconnected by links with variable strengths. A unit can be in different states according to its interaction with the units it is connected to. The strength of a connection between two units determines the degree of interaction between the two units. In many models a unit can be in only one of two discrete states. This is comparable to the firing and non-firing modes of a neuron in the human brain.

- The interaction between units and their neighbours, i.e. the units to which they are directly connected, may be *excitatory* or *inhibitory*. This is indicated by the sign of the strength of the corresponding connections. The response of a unit to its neighbours is determined by a scalar nonlinear function F of the states of the neighbours, and the corresponding connection strengths. The response takes the following general form:

$$k(u) = F \left(\sum_{v \in \mathcal{N}_u} s_{\{u,v\}}\, k(v) \right), \qquad (7.1)$$

where $k(u)$ denotes the state of unit u, $s_{\{u,v\}}$ the strength of the connection between units u and v, and $\mathcal{N}_u$ the set of neighbours of u.

- Units operate in parallel, i.e. they simultaneously try to adjust their states to the states of their neighbours. After some time the individual units may settle in a fixed state and the network then stabilizes in a global configuration. In this way, the units in the network cooperatively try to optimize a given global quantity, by using only locally available information, which enables the efficient use of massive parallelism.

- Information in connectionist models is totally distributed over the network and stored as connection strengths.

In addition to the global similarities discussed above, a number of differences can be identified among the existing models. Here, we mention the following ones.

- **Binary versus Continuous Values**

  The states of the units may be either restricted to binary values, e.g. $k(u) \in \{0, 1\}$, or to continuous values, e.g. $k(u) \in [0, 1]$. Examples of models with binary-valued units are the *perceptron model* [Rosenblatt, 1962; Minsky & Papert, 1969], Hopfield's *content-addressable memory* [Hopfield, 1982; 1984], the *backpropagation network* [Rumelhart, Hinton & Williams, 1986], and the *Boltzmann machine* [Hinton & Sejnowski, 1983; Ackley, Hinton & Sejnowski, 1985]. An example of models with continuous-valued units is the *neural decision network* of Hopfield & Tank [1985].

- **Deterministic versus Probabilistic Transitions**

  The response function F of (7.1) may be either deterministic or stochastic. Examples of models using a deterministic response function are the perceptron model, Hopfield's content-addressable memory, the backpropagation network and the neural decision network. The Boltzmann machine uses a stochastic response function.

- **Unidirectional versus Bidirectional Connections**

  If a connection from unit u to v is unidirectional, then u influences v, but not *vice versa*. If the connection is bidirectional both units influence each other.

  Models using unidirectional connections usually assume the network to be acyclic, such that information flows through the network in one direction. Examples are the perceptron modeland the backpropagation network.

  Models using bidirectional connections can be subdivided into models assuming symmetric strengths, i.e. $s_{\{u,v\}} = s_{\{v,u\}}$ for all units u, v, and models allowing asymmetric strengths. The use of asymmetric strengths may result in a periodic equilibrium behaviour of the network, which may be appealing for some applications but troublesome for others. Examples of models with bidirectional connections with symmetric strengths are Hopfield's content-addressable memory, the neural decision network, and the Boltzmann machine. An example of a connectionist model using asymmetric connections is given by Derrida, Gardner & Zippelius [1987].

- **Distributed versus Local Representations**

    In a number of models, the state of an individual unit represents a specific statement about the problem the network tries to solve. The state of the unit may indicate whether the statement is true or false. In this case the representation is called local. For example, if a network is used to parse a given sentence, then a node in the network might represent the statement that the first word of the sentence is a verb.

    In other models, the state of a single unit can only be interpreted as part of a representation given by an activation pattern over a (large) number of units. In this case the representation is called distributed.

    Whether representations are local or distributed depends on the type of application. We return to this later in the context of learning.

The criteria given above illustrate some important characteristics of the units and their connections in connectionist models, but they do not concern potential learning capabilities of these models.

Adaptation of connection strengths or learning in connectionist models is one of the major research items in neural computing [Thompson, 1986; Rumelhart, McClelland & the PDP Research Group, 1986]. Information in connectionist models is stored in the strengths of the connections. For many connectionist models, learning algorithms have been developed that are based on the hypothesis that adaptive adjustment of the 'strengths' of inter-neural connections (synapses) is what causes learning in the neural network of the human brain. This hypothesis was first proposed by Hebb in 1949, and recent neurophysiological research seems to support this hypothesis; see e.g. [Bliss & Lomo, 1973].

With respect to learning in connectionist models, we distinguish the following characteristic features.

- **Supervised versus Unsupervised Learning**

    Connectionist models that are used as classifiers train with supervision, i.e. during the training of the network, examples of objects that have to be classified are presented as inputs to the network together with a classification label that specifies the correct corresponding output.

    Connectionist models that are used as associative or content-addressable memories train without supervision. These models 'learn' by capturing the regularities in the stimuli they receive from the 'outside world', and adjust their internal connection strengths accordingly.

To achieve this, the states of a subset of the units are determined by examples of the 'outside world'. This is comparable to the firing or non-firing states of the neurons in the retina of the human eye responding to the outside world. Next, by iterative adjustment of the connection strengths, the network captures the regularities in the learning examples. In this way, the network builds up an internal model of its 'environment', which may result in memory and associative capabilities.

Examples of models using supervised learning are the perceptron model and the backpropagation network. An example of a model using unsupervised learning is the model of feature-maps [Kohonen, 1988]. The Boltzmann machine can be used for both types of learning; see Chapter 11.

- With versus Without Hidden Units

A given desired behaviour of a network, defined on a given number of units, the so-called visible or environmental units, can sometimes be learned by simply adjusting the strengths of the connections joining these units.

In many cases, however, it is necessary to extend the number of units with so-called hidden units, together with extra connections; see for instance the exclusive-or problem of Example 10.2. Hidden units are necessary to capture higher-order regularities in a given behaviour that cannot be expressed by simple co-occurrences of the states of the environmental units. Hidden units enable the construction of an internal representation. The states of hidden units are not specified directly by the given behaviour.

If no hidden units are necessary or if the number of hidden units is chosen such that the internal representation is local, then each connection strength typically corresponds to a meaningful relation between two units. Consequently, the connection strengths can be chosen by hand or by a simple learning algorithm that is based on a complete specification of the desired states of all units in the network, such as the Hebb rule [Hebb, 1949].

A local internal representation usually requires an excessive number of hidden units. Alternatively, one may use a distributed internal representation. Clearly, this makes learning much harder. Networks using a distributed internal representation require substantially less hidden units and have interesting properties, such as associative capabilities and fault tolerance [Rumelhart, McClelland & the PDP Research Group, 1986].

Examples of models without hidden units are the perceptron model and Hopfield's content-addressable memory. Examples of models with hidden units are the backpropagation network and the Boltzmann machine.

7.3   A Historical Overview

Research on neural computing has a relatively long history. The first researchers to conceive the fundamentals of neural computing were McCulloch and Pitts [1943] and Hebb [1949]. In the 1960s neural computing began to attract much attention and many models were proposed in this era. For a bibliography of the work in this decade the reader is referred to Posch [1968]. This revival was predominantly due to the work of Rosenblatt [1962] on the perceptron model.

The perceptron model can solve a number of simple classification problems using a network of binary threshold units, connected by unidirectional links. The units are divided over a number of layers and they are connected such that information can flow only from units in a higher (input) layer to units in a lower (output) layer. A learning algorithm, the so-called perceptron convergence procedure, was developed for a two-layered perceptron [Rosenblatt, 1962]. The learning algorithm indicates how the strengths of the connections can be adjusted in case the network does not give a correct answer based on the present set of connection strengths. Furthermore, Rosenblatt developed the *perceptron convergence theorem*, which states that, if there exists a proper set of connection strengths for a given classification problem, then the learning algorithm will find this set within a finite amount of time.

In 1968 Minsky and Papert demonstrated the fundamental limitations of the perceptron model. They showed that (i) for many classification problems no proper set of connection strengths exists for a two-layered network, and (ii) for many problems, for which a proper set of connection strengths does exist, the convergence of Rosenblatt's learning algorithm is very slow. These results caused a rapid decrease of the scientific interest in the subject.

Recent work, however, again shows a revival of neural computing. Within this context we mention explicitly the work of Ackley, Hinton & Sejnowski [1985], Amari & Arbib [1982], Feldman & Ballard [1982], Fukushima [1975; 1980], Grossberg [1982; 1988b], Hinton & Sejnowski [1983], Hopfield [1982; 1984], Hopfield & Tank [1985], Kohonen [1987;1988], and Rumelhart, McClelland & the PDP Research Group [1986]. This new interest is due to a number of developments.

- The recent models overcome some of the limitations that are attributed to the perceptron model.

- The continuing progress in the design and fabrication of integrated circuits enables hardware implementations of neural networks with hundreds or thousands of neurons [Alspector & Allen, 1987; Graf, Jackel, Howard, Staughn, Denker, Hubbard, Tennant & Schwartz, 1986; Mead, 1988; Sage, Thompson & Withers, 1986]. At the same time, research on opto-electronics indicates that in future dense interconnection networks will be possible by using opto-electronic devices, which apply holographic representations [Abu-Mostafa & Pslatis, 1987; Hecht-Nielsen, 1987; Ticknor & Barrett, 1987]. For an overview of parallel architectures for neural computing the reader is referred to Treleaven [1988].

- Research on artificial intelligence and pattern recognition has shown that many intelligent tasks turn out to be very hard to solve by conventional computers [Devijver & Kittler, 1987; Feldman & Ballard, 1982]. For these disciplines, massive parallelism and self-organisation may be essential for obtaining satisfactory results in a reasonable amount of time [Fahlman& Hinton, 1987; Fahlman, Hinton & Sejnowski, 1983]. Promising demonstrations in the field of speech processing are given by Kohonen, Torkkola, Shozakai, Kangas & Ventä [1987], Prager, Harrison & Fallside [1986], Sejnowski & Rosenberg [1987], and Tank & Hopfield [1987].

Despite the fact that neural computing is presently attracting much attention from both the scientific and industrial community the subject is still in its infancy. However, the current developments hold many a promise for the future.

7.4   The Boltzmann Machine

The Boltzmann machine, introduced by Hinton & Sejnowski [1983], is a recent approach to connectionist models. The model can be viewed as a generalization of Hopfield's content-addressable memory. However, the basic idea underlying the Boltzmann machine, i.e. the implementation of local constraints as connection strengths in stochastic networks, was introduced by Moussouris [1974].

As in Hopfield's model, the units in a Boltzmann machine have binary-valued states, i.e. they are either 'on' or 'off', and the connections are bidirectional. The Boltzmann machine uses a probabilistic state transition mechanism, while Hopfield's model uses a deterministic one. Furthermore, a Boltzmann machine can have hidden units to capture higher-order regularities when learning.

The strength of a connection in a Boltzmann machine can be considered as a quantitative measure of the desirability that the units joined by the connection are both 'on'. The units in a Boltzmann machine try to reach a maximal consensus about their individual states, subject to the desirabilities expressed by the connection strengths. To adjust the states of the individual units to the states of their neighbours a probabilistic state transition mechanism is used which is governed by the simulated annealing algorithm.

There are three main reasons for studying the model of Boltzmann machines:

- The model offers a generalized approach that can be applied to all three basic research issues in the field of neural computing, viz. *search, representation* and *learning* [Hinton, 1987].

- The model is supported by a rigorous mathematical formalism that provides a good description of the convergence properties of the network and enables the construction of a simple learning algorithm, suited for supervised as well as unsupervised learning.

- Due to its simplicity the model is relatively easy to put directly onto silicon.

The remaining chapters of this book are devoted to a detailed description of the Boltzmann machine. The organization of the chapters is as follows.

In Chapter 8 we present a mathematical formalism of the Boltzmann machine. The formalism contains a description of its structure and the stochastic state transition mechanism that is applied to maximize the consensus.

The mathematical model of the Boltzmann machine has two emergent features. On the one hand, the Boltzmann machine can be considered as a model for a massively parallel implementation of the simulated annealing algorithm which can be used to solve (approximately) combinatorial optimization problems. On the other hand, the introduction of simulated annealing into the field of neural computing enables the design of a learning algorithm, which is based on relatively simple concepts, leading to some interesting properties.

In Chapter 9 we discuss the relation between the Boltzmann machine and combinatorial optimization. A general procedure is presented for solving combinatorial optimization problems within the model of the Boltzmann machine. The use of this procedure is illustrated by a number of examples.

As an intermediate between combinatorial optimization and learning, we discuss in Chapter 10 the relation between classification and the model of Boltzmann machines. By means of two simple examples we illustrate the different aspects that are related to the use of Boltzmann machines for solving classification problems.

The subject of learning in a Boltzmann machine is addressed in Chapter 11. This chapter presents a detailed derivation of a learning algorithm for Boltzmann machines based on the mathematical formalism presented in Chapter 8. Furthermore, a discussion is presented on the practical aspects of learning, reviewing a number of applications of the Boltzmann machine in this area.

CHAPTER 8

Boltzmann Machines

In this chapter we present a formal description of the Boltzmann machine based on a graph-theoretical formulation of the architectural structure of the underlying network; see Section 8.1. Furthermore, a detailed description is given of the stochastic state transition mechanism - based on the simulated annealing algorithm - that is applied to maximize the consensus in a Boltzmann machine. For this description we distinguish between sequential and parallel Boltzmann machines; see Sections 8.2 and 8.3, respectively. The chapter ends with a taxonomy that surveys the possible application areas of the Boltzmann machine.

8.1 Structural Description

A Boltzmann machine can be viewed as a network consisting of a number of two-state units, that are connected in some way. The network is represented by a pseudograph $B = (\mathcal{U}, C)$, where $\mathcal{U}$ denotes the finite set of units and C is a set of unordered pairs of elements of $\mathcal{U}$ denoting the connections between the units. The set of connections usually includes all *loops* or *bias connections*, i.e. $\{\{u, u\} \mid u \in \mathcal{U}\} \subset C$. If two units are connected, they are called *adjacent*. Furthermore, we say that the set C specifies the *connection pattern* of a Boltzmann machine.

A *unit* u can be in one of two states: a unit is either 'on' or 'off'. With each unit, a 0-1 variable is associated, denoting the state of the unit, where 0 and 1 correspond to 'off' and 'on', respectively. A *connection* $\{u, v\} \in C$ joins the units u and v.

Definition 8.1 A *configuration* k of a Boltzmann machine is given by a global state of the Boltzmann machine and is uniquely defined by a sequence of length $|\mathcal{U}|$, whose u^{th} component $k(u)$ denotes the *state* of unit u in configuration k. The *configuration space* $\mathcal{R}$ is given by the set of all possible configurations. Clearly, the cardinality of $\mathcal{R}$ equals $2^{|\mathcal{U}|}$. ∎

Definition 8.2 Let $\{u, v\}$ be a connection joining the units u and v, then $\{u, v\}$ is *activated* in a given configuration k if both units u and v are 'on', i.e. if

$$k(u) \cdot k(v) = 1. \tag{8.1}$$

Otherwise the connection is not activated. ∎

Definition 8.3 With a connection $\{u, v\} \in C$ a *connection strength* $s_{\{u,v\}} \in \mathbb{R}$ is associated. The connection strength $s_{\{u,v\}}$ is a quantitative measure for the *desirability* that connection $\{u, v\}$ is activated. By definition, $s_{\{u,v\}} = s_{\{v,u\}}$. If $s_{\{u,v\}} > 0$ it is desirable that $\{u, v\}$ is activated; if $s_{\{u,v\}} < 0$ it is undesirable. Connections with a positive strength are called *excitatory*; connections with a negative strength are called *inhibitory*. Furthermore, the strength $s_{\{u,u\}}$ of the bias connection $\{u, u\}$ is called the *bias* of unit u. ∎

Since the connection strengths impose local constraints, Boltzmann machines are also often called *constraint satisfaction networks* [Hinton, Sejnowki & Ackley, 1984].

As an overall desirability measure of the individual states of the units in a Boltzmann machine, we introduce the *consensus* as the value of the consensus function which is defined as follows.

Definition 8.4 The *consensus function* $C : \mathcal{R} \rightarrow \mathbb{R}$ assigns to each configuration k a real number that is given by the sum of the strengths of the activated connections, i.e.

$$C(k) = \sum_{\{u,v\} \in C} s_{\{u,v\}} \, k(u) \, k(v). \tag{8.2}$$

  ∎

The consensus is large if many excitatory connections are activated, and it is small if many inhibitory connections are activated. In fact, the consensus is a global measure indicating to what extent the units in a Boltzmann machine have reached a consensus about their individual states, subject to the desirabilities expressed by the individual connection strengths.

The objective of a Boltzmann machine is to reach a globally maximal configuration, i.e. a configuration with maximal consensus. To reach a maximal consensus, a *state transition mechanism* is introduced which allows the units to adjust their states to those of their neighbours. The adjustment is determined by a stochastic function of the states of the neighbours and the corresponding connection strengths. Hence, as in the simulated annealing algorithm, the state transition mechanism uses a stochastic acceptance criterion, allowing a Boltzmann machine to escape from locally optimal configurations.

To describe the consensus maximization in a Boltzmann machine, we distinguish between the following two models:

- *sequential Boltzmann machines*, in which units are allowed to change their states only one at a time, and

- *parallel Boltzmann machines*, in which units are allowed to change their states simultaneously.

In the following two sections we elaborate on these two models in more detail.

8.2 Sequential Boltzmann Machines

Essential to a sequential Boltzmann machine is that units may change their states only one at a time.

Definition 8.5 Let a Boltzmann machine be in a configuration k, then a *neighbouring configuration* k_u is defined as the configuration that is obtained from k by changing the state of unit u (from 0 to 1 or vice versa). Hence, we have

$$k_u(v) = \begin{cases} k(v) & \text{if } v \neq u \\ 1 - k(v) & \text{if } v = u. \end{cases} \tag{8.3}$$

The *neighbourhood* $R_k \subset R$ is defined as the set of all neighbouring configurations of k. ∎

Let C_u denote the set of connections incident with unit u, excluding $\{u, u\}$, and let $C' = C - C_u - \{u, u\}$. The *difference in consensus* between the configurations k and k_u is denoted by

$$\Delta C_k(u) = C(k_u) - C(k) \tag{8.4}$$

Realizing that the contribution of the connections in C' to the consensus is identical for both k and k_u, we obtain

$$\Delta C_k(u) = \left[k_u(u) \sum_{\{u,v\} \in C_u} s_{\{u,v\}} \, k_u(v) + k_u^2(u) s_{\{u,u\}} \right]$$
$$- \left[k(u) \sum_{\{u,v\} \in C_u} s_{\{u,v\}} \, k(v) + k^2(u) s_{\{u,u\}} \right]. \tag{8.5}$$

Using (8.3) we then obtain

$$\Delta C_k(u) = (1 - 2\,k(u)) \left[\sum_{\{u,v\}\in C_u} s_{\{u,v\}}\, k(v) + s_{\{u,u\}} \right]. \tag{8.6}$$

From (8.6) it is apparent that the effect on the consensus, resulting from changing the state of unit u, is completely determined by the states of the neighbours of u and the corresponding connection strengths. Consequently, each unit can evaluate locally its own state transition since no global calculations are required. This is a very important property since it means that there is potential for parallel execution; see Section 8.2.2.

Definition 8.6 Let the neighbourhoods be given by Definition 8.5, then a *locally maximal configuration*, or simply *local maximum*, is defined as a configuration $k \in R$ where

$$\Delta C_k(u) \le 0 \quad \text{for all} \quad u \in \mathcal{U}. \tag{8.7}$$

Thus, a local maximum is a configuration whose consensus cannot be increased by a single state transition. The set of all local maxima is denoted by $\hat{R}$.   ∎

As in the simulated annealing algorithm, the concept of Markov chains can be used to describe the state transitions of the units in a Boltzmann machine; see Chapter 3. Again we need to define trials and a transition probability. In a sequential Boltzmann machine a trial consists of two steps. Given a configuration k, then firstly, a unit u is selected which will then try to change its state, i.e. a neighbouring configuration k_u is generated, and secondly, it is evaluated whether or not the configuration k_u is accepted. Furthermore, a control parameter c is introduced that determines the probability of accepting state transitions. Hence, the theory of Markov chains now can be used as follows.

Let $\mathbf{X}(m)$ denote the outcome of the mth trial and let $k, l \in R$ be a pair of configurations, then the transition probability $P_{kl}(c)$ of transforming configuration k into l is defined by (cf. Definition 3.5)

$$P_{kl}(c) = \mathbb{P}_c\{\mathbf{X}(m) = l \mid \mathbf{X}(m - 1) = k\}, \tag{8.8}$$

and

$$P_{kl}(c) = \begin{cases} G(u)A_k(u, c) & \text{if } l = k_u \\ 1 - \sum_{u\in\mathcal{U}} P_{kk_u}(c) & \text{if } l = k \\ 0 & \text{otherwise,} \end{cases} \tag{8.9}$$

where $G(u)$ denotes the probability of generating a state transition of unit u, and $A_k(u, c)$ denotes the probability of accepting the state transition of unit u given configuration k; c denotes the control parameter ($c \in \mathbb{R}^+$).

The generation probability is usually chosen uniformly over the available units and independent of the configuration k and the control parameter c. This choice is motivated primarily by the fact that we would like the generation probability to be as simple as possible. All units in a Boltzmann machine

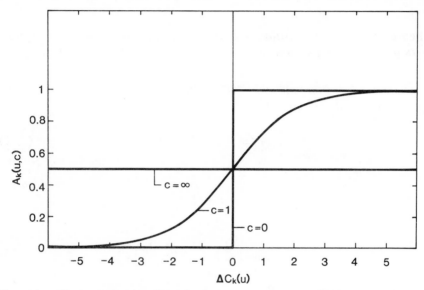

Figure 8.1: The acceptance probability $A_k(u, c)$ as a function of $\Delta C_k(u)$ for different values of the control parameter c.

are considered to be identical and therefore there is no evidence beforehand for treating some units differently with respect to generating state transitions. The acceptance probability is chosen as

$$A_k(u, c) = \frac{1}{1 + \exp\left(\frac{-\Delta C_k(u)}{c}\right)}, \qquad (8.10)$$

where $\Delta C_k(u)$ is given by (8.6). Figure 8.1 shows the acceptance probability $A_k(u, c)$ as a function of $\Delta C_k(u)$ for different values of the control parameter c. The expression of (8.10) differs from the standard choice used in the simulated annealing algorithm, as given by (3.8). The difference is only minor. Both acceptance probabilities lead to the same stationary distribution and thus have the same convergence properties; cf. Theorems 3.2 and 8.1. The expression of (8.10) is merely chosen so as to reflect the typical sigmoid response of neurons in a (biological) neural network. Moreover, the symmetric shape of (8.10) is much easier to realize in hardware. It thus is considered a natural choice for the Boltzmann machine.

We now can prove that the Boltzmann machine converges asymptotically to the set of globally optimal configurations. The proof of this is quite like the proof of the asymptotic convergence of the simulated annealing algorithm, as given in Theorem 3.2, and is based on the existence of a stationary distribution.

Theorem 8.1 *Let the transition probabilities in a Boltzmann machine be given by (8.9), where the acceptance probabilities are given by (8.10), then*

  (*i*) *there exists a unique stationary distribution* $\mathbf{q}(c)$ *for all* $c > 0$, *whose components are given by*

$$q_k(c) \;=\; \lim_{m\to\infty} \mathbb{P}_c\{X(m) = k\} \tag{8.11}$$

$$\;=\; \frac{1}{N_0(c)} \exp\left(\frac{C(k)}{c}\right), \tag{8.12}$$

where

$$N_0(c) = \sum_{l\in\mathcal{R}} \exp\left(\frac{C(l)}{c}\right), \tag{8.13}$$

and

  (*ii*) *the stationary distribution converges (as* $c \downarrow 0$*) to a uniform distribution over the set of optimal configurations, i.e.*

$$\lim_{c\downarrow 0}\lim_{m\to\infty} \mathbb{P}_c\{X(m) = k\} = \frac{1}{|\mathcal{R}_{opt}|}\mathcal{X}_{(\mathcal{R}_{opt})}(k), \tag{8.14}$$

where $\mathcal{R}_{opt}$ *denotes the set of optimal configurations.*

Proof From the choice of the generation probabilities it follows directly that each configuration can be generated from any other configuration in a finite number of steps. Hence, condition (1) of Theorem 3.2 is satisfied. Following the lines of the proof of Theorem 3.2, it then follows directly that the Markov chain induced by the transition probabilities of (8.9) is irreducible and aperiodic. We now use Lemma 3.2 to complete the proof of part (*i*) of the theorem, i.e. we prove that $q_k(c)P_{kl}(c) = q_l(c)P_{lk}(c)$ for all $k, l \in \mathcal{R}$.

Let $l \in \mathcal{R}_k$ (otherwise the proof is trivial), then there exists a unit u such that $l = k_u$. Hence,

$$q_k(c)P_{kl}(c) \quad = \quad q_k(c)G(u)A_k(u,c)$$

$$= \quad \frac{1}{N_0(c)}\exp\left(\frac{C(k)}{c}\right)G(u)\frac{1}{1+\exp\left(\frac{-\Delta C_k(u)}{c}\right)}$$

$$= \quad \frac{1}{N_0(c)}\exp\left(\frac{C(k_u)}{c}\right)G(u)\frac{\exp\left(\frac{-(C_{k_u}-C_k)}{c}\right)}{1+\exp\left(\frac{-\Delta C_k(u)}{c}\right)}$$

$$= \quad \frac{1}{N_0(c)}\exp\left(\frac{C(k_u)}{c}\right)G(u)\frac{1}{1+\exp\left(\frac{-\Delta C_{k_u}(u)}{c}\right)}$$

$$= \quad q_{k_u}(c)G(u)A_{k_u}(u,c) = q_l(c)P_{lk}(c).$$

The proof of part (ii) of the theorem is equivalent to the proof of Corollary 2.1.

∎

As in the simulated annealing algorithm, a finite-time approximation is obtained by specifying a set of parameters that determine a cooling schedule; see Chapter 4. In a finite-time approximation, the Boltzmann machine starts off at a sufficiently large initial value of c, and a randomly chosen initial configuration. Subsequently, a sequence of Markov chains of finite length is generated at descending values of c. Eventually, as c approaches 0, state transitions become more and more infrequent, and finally the Boltzmann machine stabilizes in a locally maximal configuration which is taken as the final configuration. It should be noted that the final configuration will be a (near-)optimal one, since, as in the simulated annealing algorithm, the finite-time approximation no longer guarantees convergence to an optimal configuration.

Finally, we mention that the cooling schedules presented in Chapter 4 can also be applied to obtain a finite-time implementation of a sequential Boltzmann machine.

To illustrate some of the concepts presented in this section, we discuss the following example.

Example 8.1 Let $\mathcal{B} = (\mathcal{U}, \mathcal{C})$ be a Boltzmann machine with $\mathcal{U} = \{u_1, \ldots, u_9\}$ the set of units and $\mathcal{C} = \{\{u_j, u_j\} \mid u_j \in \mathcal{U}\} \cup \mathcal{C}_i$ the set of connections, where

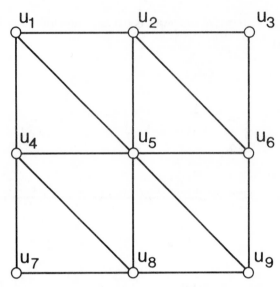

Figure 8.2: An example of a small Boltzmann machine. The bias connections are not drawn.

C_i denotes the set of connections shown in Figure 8.2. The bias connections are given a strength equal to 1; other connections are given a strength equal to -2. As all units have a positive bias, they will all have a tendency to be 'on'. However, the negative strengths of the other connections have the effect that eventually, after the optimization of the consensus, no two adjacent units will be 'on' at the same time. It is easy to verify that in a configuration $k \in \hat{R}$ only bias connections are activated. Since, if in a given configuration k a connection $\{u, v\}$, with $u \neq v$, is activated, then changing the state of unit u (or v) raises the consensus with $\Delta C_k(u) \geq 1$. Clearly, the configuration for which the consensus is maximal is given by the configuration $k \in \hat{R}$ for which the number of activated bias connections is maximal. This is the case for the configuration where units u_1, u_3, u_7, and u_9 are 'on' and other units are 'off'. Even for this small Boltzmann machine there are 12 local maxima of the consensus function.   ∎

8.3   Parallel Boltzmann Machines

Following the principles of connectionism, one strives for parallel execution of state transitions. Boltzmann machines facilitate the use of parallelism since units can evaluate their state transitions locally. Furthermore, due to the distributed representation of a configuration, a Boltzmann machine does not suffer from the Von Neumann bottleneck as in more traditional computer architectures.

There are a number of different approaches that can be pursued to realize parallel state transitions in a Boltzmann machine. Here, we distinguish between the following two modes of parallelism.

- *Synchronous parallelism.* Sets of state transitions are scheduled in successive trials, each trial consisting of a number of individual state transitions. After each trial the accepted state transitions are communicated through the network so that all units have up-to-date information about the states of their neighbours before the next trial is initiated. During each trial a unit is allowed to propose a state transition exactly once. Evidently, synchronous parallelism requires a global clocking scheme to control synchronization.

- *Asynchronous parallelism.* State transitions are evaluated simultaneously and independently. Units continuously generate state transitions and accept or reject them on the basis of information that is not necessarily up-to-date, since the states of its neighbours may have changed in the meantime. Clearly, asynchronous parallelism does **not** require a global clocking scheme, which is of great advantage in hardware implementations of the Boltzmann machine.

In the following two sections we discuss these modes of parallelism in greater detail.

8.3.1 Synchronous Parallelism

A trial in a synchronously parallel Boltzmann machine consists of two steps. Firstly, a subset $\mathcal{U}_s \subset \mathcal{U}$ of units is generated that may propose a state transition, and secondly, each unit in $\mathcal{U}_s$ evaluates whether or not to accept a state transition. All other units remain unchanged. Thus, if we define for an arbitrary pair of configurations $k, l \in \mathcal{R}$ the set $\mathcal{U}_{kl}$ as the subset of units that must change their states in order to transform k into l, i.e.

$$\mathcal{U}_{kl} = \{u \in \mathcal{U} \mid k(u) \neq l(u)\}, \tag{8.15}$$

(note that $\mathcal{U}_{kl} = \mathcal{U}_{lk}$), then we have the following result. If the current configuration is k, then the next configuration is l if (i) $\mathcal{U}_s \supseteq \mathcal{U}_{kl}$ and (ii) all units in $\mathcal{U}_{kl}$ accept a state transition and all units in $\mathcal{U}_s \backslash \mathcal{U}_{kl}$ reject a state transition. Consequently, the probability of such a transition is given by

$$P_{kl}(c) \quad = \quad \mathbb{P}_c\{\mathbf{X}(m) = l \mid \mathbf{X}(m-1) = k\}, \tag{8.16}$$

$$= \begin{cases} \displaystyle\sum_{\mathcal{U}_s \supseteq \mathcal{U}_{kl}} G(\mathcal{U}_s) \prod_{u \in \mathcal{U}_{kl}} A_k(u, c) \prod_{v \in \mathcal{U}_s \backslash \mathcal{U}_{kl}} R_k(v, c) & \text{if } l \neq k \\[2em] 1 - \displaystyle\sum_{m \in \mathcal{R}, m \neq k} P_{km}(c) & \text{if } l = k, \end{cases}$$

$$\tag{8.17}$$

where the stochastic variable $\mathbf{X}(m)$ denotes the outcome of the mth trial, $G(\mathcal{U}_s)$ denotes the probability of generating the subset $\mathcal{U}_s$ and $R_k(u, c)$ denotes the rejection probability given by $R_k(u, k) = 1 - A_k(u, c)$.

We now distinguish between the following two cases.

- *Limited parallelism:* units may change their states in parallel only if they are not adjacent.

- *Unlimited parallelism:* units may change their states in parallel whether or not they are adjacent.

Below we discuss the convergence properties of synchronously parallel Boltzmann machines. As will become apparent, these properties are markedly different for unlimited parallelism as compared to limited parallelism.

Later on we will show that asymptotic convergence can be proved for synchronously parallel Boltzmann machines applying limited parallelism. Furthermore, we will indicate where the proof fails if unlimited parallelism is applied.

Limited Parallelism

To prove asymptotic convergence of a synchronously parallel Boltzmann machine applying limited parallelism, we consider the special case where the set of units $\mathcal{U}$ is partitioned into m independent sets $\mathcal{U}_s$ (see also Example 8.2), i.e.

$$\mathcal{U} = \bigcup_{s=1}^{m} \mathcal{U}_s, \tag{8.18}$$

$$\mathcal{U}_s \cap \mathcal{U}_p = \emptyset, \quad s \neq p = 1, \ldots, m, \quad \text{and} \tag{8.19}$$

$$\forall u, v \in \mathcal{U}_s : \{u, v\} \notin C, \quad s = 1, \ldots, m. \tag{8.20}$$

Clearly, (8.20) implies that units belonging to the same subset $\mathcal{U}_s$ are not adjacent. Thus, limited parallelism is effected by allowing units, belonging to the same subset $\mathcal{U}_s$, to change their states simultaneously.

As a consequence of the conditions given in (8.18) - (8.20), we obtain the following results. Let $k, l \in \mathcal{R}$ be two configurations and let $\mathcal{U}_s$ be an independent set such that $\mathcal{U}_{kl} \subseteq \mathcal{U}_s$. Then it is straightforward to show the correctness of the following relations:

$$C(l) - C(k) = \sum_{u \in \mathcal{U}_{kl}} \Delta C_k(u), \tag{8.21}$$

and

$$\Delta C_k(u) = \begin{cases} -\Delta C_l(u) & \text{if } u \in \mathcal{U}_{kl} \\ \Delta C_l(u) & \text{if } u \in \mathcal{U}_s \backslash \mathcal{U}_{kl}. \end{cases} \tag{8.22}$$

It is important to note that the relations (8.21) and (8.22) do not hold if the set $\mathcal{U}_s$ is not independent, i.e. if $\mathcal{U}_s$ contains units that are mutually connected. We return to this later.

We now can prove asymptotic convergence for the synchronously parallel Boltzmann machine, using limited parallelism. This is formulated in the following theorem.

Theorem 8.2 *Let the set of units be partitioned into independent sets, and let the transition probabilities in a Boltzmann machine be given by (8.17), where the acceptance probabilities are given by (8.10), and let the generation probability be uniform over the number of independent sets, then*

(*i*) *there exists a unique stationary distribution* $\mathbf{q}(c)$, *whose components are given by*

$$q_k(c) = \lim_{m \to \infty} \mathbb{P}_c\{\mathbf{X}(m) = k\} \tag{8.23}$$

$$= \frac{1}{N_0(c)} \exp\left(\frac{C(k)}{c}\right), \tag{8.24}$$

where

$$N_0(c) = \sum_{l \in \mathcal{R}} \exp\left(\frac{C(l)}{c}\right), \tag{8.25}$$

and

(*ii*) *the stationary distribution converges (as $c \downarrow 0$) to a uniform distribution over the set of optimal configurations, i.e. we have*

$$\lim_{c \downarrow 0} \lim_{m \to \infty} \mathbb{P}_c\{\mathbf{X}(m) = k\} = \frac{1}{|\mathcal{R}_{opt}|} \chi_{(\mathcal{R}_{opt})}(k), \tag{8.26}$$

where $\mathcal{R}_{opt}$ denotes the set of optimal configurations.

Proof The proof is similar to the proof of Theorem 8.1. Again, it is easily verified that the Markov chain induced by the transition probabilities of (8.17) is irreducible and aperiodic. The proof is completed by showing the validity of the detailed balance equation, i.e. for all $k, l \in \mathcal{R}$: $q_k(c)P_{kl}(c) = q_l(c)P_{lk}(c)$.

Let $l \neq k$ and $\mathcal{U}_{kl} \subseteq \mathcal{U}_s$, otherwise the proof is trivial. Note that the set $\mathcal{U}_s$ is unique. Then by using (8.17), (8.21) and (8.22) we obtain

$$
\begin{aligned}
q_k(c)P_{kl}(c) &= q_k(c)\, G(\mathcal{U}_s) \prod_{u \in \mathcal{U}_{kl}} A_k(u, c) \prod_{v \in \mathcal{U}_s \backslash \mathcal{U}_{kl}} R_k(v, c) \\[3mm]
&= \frac{1}{N_0(c)} \exp\left(\frac{C(k)}{c}\right) G(\mathcal{U}_s) \prod_{u \in \mathcal{U}_{kl}} A_k(u, c) \prod_{v \in \mathcal{U}_s \backslash \mathcal{U}_{kl}} R_k(v, c) \\[3mm]
&= \frac{1}{N_0(c)} \exp\left(\frac{C(l)}{c}\right) \exp\left(\frac{-(C(l) - C(k))}{c}\right) \\
&\qquad \times G(\mathcal{U}_s) \prod_{u \in \mathcal{U}_{kl}} A_k(u, c) \prod_{v \in \mathcal{U}_s \backslash \mathcal{U}_{kl}} R_k(v, c) \\[3mm]
&= q_l(c) \exp\left(\frac{\sum\limits_{u \in \mathcal{U}_{kl}} -\Delta C_k(u)}{c}\right) \\
&\qquad \times G(\mathcal{U}_s) \prod_{u \in \mathcal{U}_{kl}} \frac{1}{1 + \exp\left(\frac{-\Delta C_k(u)}{c}\right)} \prod_{v \in \mathcal{U}_s \backslash \mathcal{U}_{kl}} R_k(v, c) \\[3mm]
&= q_l(c) G(\mathcal{U}_s) \prod_{u \in \mathcal{U}_{kl}} \frac{\exp\left(\frac{-\Delta C_k(u)}{c}\right)}{1 + \exp\left(\frac{-\Delta C_k(u)}{c}\right)} \prod_{v \in \mathcal{U}_s \backslash \mathcal{U}_{kl}} R_k(v, c) \\[3mm]
&= q_l(c)\, G(\mathcal{U}_s) \prod_{u \in \mathcal{U}_{lk}} A_l(u, c) \prod_{v \in \mathcal{U}_s \backslash \mathcal{U}_{lk}} R_l(v, c)
\end{aligned}
$$

$$= \quad q_l(c) \, P_{lk}(c).$$

Again, the proof of part (ii) is equivalent to the proof of Corollary 2.1. ∎

Limited parallelism can be exploited to the full if the set of units is partitioned into a minimal number of independent sets. The maximal speed-up that can be achieved in this case equals $\frac{|\mathcal{U}|}{m_{min}}$, where m_{min} denotes the minimal number of independent sets into which $\mathcal{U}$ can be partitioned.

We mention that the problem of partitioning $\mathcal{U}$ into a minimal number of independent sets is equivalent to the graph colouring problem; see Sections 5.1.3 and 5.1.4. Hence, the minimal number of independent sets for a given Boltzmann machine can be determined by solving the corresponding graph colouring problem. However, we recall that this problem is NP-complete, which may necessitate the use of approximation algorithms for large Boltzmann machines. On the other hand, most applications use a very regular connection pattern of the Boltzmann machine (cf. Chapters 9, 10, and 11), for which it is usually easy to find the minimum number of independent sets by hand.

For densely connected Boltzmann machines the minimal number of independent sets will be large, resulting in a relatively small speed-up. The speed-up can be increased by dropping the requirement that the set of independent sets is a partition of $\mathcal{U}$, i.e. by dropping constraint (8.19). Using a conjunction of independent sets, instead of a partition, amounts to enlarging the independent sets to contain a maximal number of units, while the sets remain independent. Thus, a unit may belong to more than one subset. This increases the probability of units being selected to propose a state transition. Enlarging the independent sets is only a minor modification, which basically does not affect the asymptotic convergence properties of the Boltzmann machine. Furthermore, in most cases there will only be a slight increase in the speed-up, since the number of units belonging to two or more independent sets is usually small.

Example 8.2 Let $\mathcal{B} = (\mathcal{U}, \mathcal{C})$ be a Boltzmann machine with $\mathcal{U} = \{u_1, \ldots, u_{25}\}$ the set of units and $\mathcal{C}$ the set of connections shown in Figure 8.3. It is easy to verify that a unique minimal partition of $\mathcal{U}$ exists consisting of the following three independent sets:

$$\mathcal{U}_1 = \{u_1, u_4, u_8, u_{12}, u_{15}, u_{16}, u_{19}, u_{23}\},$$
$$\mathcal{U}_2 = \{u_2, u_5, u_6, u_9, u_{13}, u_{17}, u_{20}, u_{21}, u_{24}\}, \text{ and}$$
$$\mathcal{U}_3 = \{u_3, u_7, u_{10}, u_{11}, u_{14}, u_{18}, u_{22}, u_{25}\}.$$

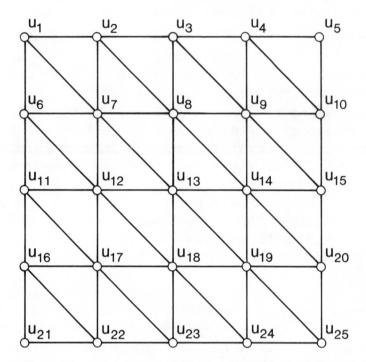

Figure 8.3: An example of a Boltzmann machine with 25 units.

Enlarging the independent sets and still keeping them independent is not possible. The maximal speed-up for the limited form of parallelism equals $\frac{|U|}{3}$ = $8\frac{1}{3}$. ∎

Unlimited Parallelism

In a synchronously parallel Boltzmann machine, applying unlimited parallelism, neighbours are allowed to change their states simultaneously. This may lead to erroneously calculated differences in consensus as follows. Suppose that two neighbours u and v simultaneously accept a state transition. This may result in the activation or deactivation of the connection $\{u, v\}$ which is not accounted for in the evaluation of the differences in consensus $\Delta C_k(u)$ and $\Delta C_k(v)$. For example, if both units u and v are 'off' and they both accept a state transition to 'on' (based on the information that they are both 'off'), then connection $\{u, v\}$ is activated, whereas it is not accounted for in the calculation of the difference in consensus and therefore not in the acceptance criterion of (8.10). Evidently, the same holds for state transitions in the reverse order. This may lead to a decrease of the consensus, despite the fact that both units may have anticipated an increase of the consensus; see also Section 6.3.

As a consequence of these erroneously calculated state transitions, the relations (8.21) and (8.22) no longer hold. Thus, in this case, asymptotic convergence can no longer be proved along the lines of the proof of Theorem 8.2, except for the special case of full parallelism, which is discussed at the end of this section. However, intuitive arguments suggests that asymptotic convergence takes place in the following way. The probability of erroneously calculated state transitions occurring decreases during the consensus optimization process as the value of the control parameter c decreases. From (8.10) it follows that the probability of a unit accepting a state transition decreases as the value of the control parameter c decreases, and eventually it approaches 0 as c approaches 0. Thus, the probability that two (or more) neighbours simultaneously change their states becomes very small in the course of the optimization process. Consequently, at small values of c, the parallel Boltzmann machine will roughly act as a sequential Boltzmann machine (with respect to accepting state transitions). We therefore conjecture that erroneously accepted state transitions do not affect the asymptotic convergence properties of the Boltzmann machine. They merely may have an effect on the speed of convergence.

The conjecture stated above is supported by a large number of computer simulations. These simulations show that the Boltzmann machine stabilizes in a configuration corresponding to (near-)optimal values of the consensus function. In our opinion, a rigorous proof of this conjecture is an interesting open research topic.

Unlimited parallelism can be maximally exploited in a Boltzmann machine by introducing some additional locality concepts. For this we leave the approach that is based on the assumption that there exists a finite number of predetermined subsets $\mathcal{U}_s$ of $\mathcal{U}$ that can be generated with a probability $G(\mathcal{U}_s)$ during each trial. Instead we use an approach that is based on the assumption that in each cycle of the global clocking scheme, used to synchronize the state transitions, each unit u may propose a state transition with a probability $G(u)$. This is similar to a sequential Boltzmann machine but with the additional degree of freedom that all units may operate simultaneously.

Thus, in each trial each unit is selected with probability $G(u)$ to propose a state transition. Hence, all possible subsets $\mathcal{U}_s \subseteq \mathcal{U}$ may occur and the generation probability $G(\mathcal{U}_s)$ is given by

$$G(\mathcal{U}_s) = \prod_{u \in \mathcal{U}_s} G(u) \prod_{v \in \mathcal{U} \backslash \mathcal{U}_s} (1 - G(v)). \qquad (8.27)$$

Hence, let $\mathcal{U}_{kl}$ again denote the set of units that have to change their states in order to transform configuration k into configuration l, then the corresponding transition probabilities of (8.17) can be rewritten as

$$P_{kl}(c) = \begin{cases} \displaystyle\prod_{u \in \mathcal{U}_{kl}} G(u) A_k(u, c) \prod_{v \in \mathcal{U} \backslash \mathcal{U}_{kl}} (1 - G(v) + G(v) R_k(v, c)) & \text{if } l \neq k \\[4mm] 1 - \displaystyle\sum_{m \in \mathcal{R}, m \neq k} P_{km}(c) & \text{if } l = k. \end{cases}$$

$$(8.28)$$

The advantage of this approach is that all calculations can be done locally, i.e. during each trial, each individual unit can locally evaluate whether or not to propose and to accept a state transition. The clocking scheme evidently remains global.

We finally discuss the special case of *full parallelism* where all units $u \in \mathcal{U}$ simultaneously propose a state transition, i.e. $G(u) = 1$ for all $u \in \mathcal{U}$. Then the transition probabilities of (8.28) can be rewritten as

$$P_{kl}(c) = \begin{cases} \displaystyle\prod_{u \in \mathcal{U}_{kl}} A_k(u, c) \prod_{v \in \mathcal{U} \backslash \mathcal{U}_{kl}} R_k(v, c) & \text{if } l \neq k \\[4mm] 1 - \displaystyle\sum_{m \in \mathcal{R}, m \neq k} P_{km}(c) & \text{if } l = k. \end{cases}$$

$$(8.29)$$

Next, by recalling that

$$A_k(u, c) = \frac{1}{1 + \exp\left(-\frac{\Delta C_k(u)}{c}\right)} \qquad (8.30)$$

and

$$R_k(u, c) = 1 - A_k(u, c) = \frac{1}{1 + \exp\left(\frac{\Delta C_k(u)}{c}\right)}, \qquad (8.31)$$

and by setting $\Delta C_k(u) = r_k(u) h_k(u)$, where

$$r_k(u) = 1 - 2k(u) \qquad (8.32)$$

and

$$h_k(u) = \sum_{\{u,v\} \in C_u} s_{\{u,v\}} k(v) + s_{\{u,u\}}, \tag{8.33}$$

the transition probabilities of (8.29) can be rewritten as

$$P_{kl}(c) = \begin{cases} \prod_{u \in \mathcal{U}} \left[1 + \exp\left(\frac{r_l(u)h_k(u)}{c} \right) \right]^{-1} & \text{if } l \neq k \\ \\ 1 - \sum_{m \in \mathcal{R}, m \neq k} P_{km}(c) & \text{if } l = k. \end{cases} \tag{8.34}$$

Note that $r_k(u) \in \{-1, 1\}$ for all $k \in \mathcal{R}$ and $u \in \mathcal{U}$. We can now state the following theorem.

Theorem 8.3 *Let the transition probabilities in a Boltzmann machine be given by (8.34), then*

(i) *there exists a unique stationary distribution $\mathbf{q}(c)$, whose components are given by*

$$q_k(c) = \frac{1}{K_0(c)} \prod_{u \in \mathcal{U}} 2\cosh\left(\frac{h_k(u)}{2c} \right) \exp\left(-\frac{h_k(u) + 2s_{\{u,u\}}k(u)}{2c} \right),$$

$$\tag{8.35}$$

where

$$K_0(c) = \sum_{l \in \mathcal{R}} \prod_{u \in \mathcal{U}} 2\cosh\left(\frac{h_l(u)}{2c} \right) \exp\left(-\frac{h_l(u) + 2s_{\{u,u\}}l(u)}{2c} \right),$$

$$\tag{8.36}$$

and

(ii) *the stationary distribution converges (as $c \downarrow 0$) to a uniform distribution over the set $\tilde{\mathcal{R}}_{opt}$, where $\tilde{\mathcal{R}}_{opt}$ denotes the set of pseudo optimal configurations, i.e. the configurations k for which the pseudo consensus function*

$$C(k) = -\sum_{u \in \mathcal{U}} \left\{ h_k(u) + 2s_{\{u,u\}}k(u) - |h_k(u)| \right\} \tag{8.37}$$

is maximal.

Proof The proof runs similarly to the proof of Theorem 8.1. The Markov chain induced by the transition probabilities of (8.34) is irreducible and aperiodic. This is a direct consequence of the choice of the generation probabilities $G(u)$. Hence, part (i) can be proved by showing that the detailed balance equation is satisfied, i.e. let $k \neq l$, we then prove that

$$q_k(c) \frac{P_{kl}(c)}{P_{lk}(c)} = q_l(c).$$

The proof runs as follows:

$$\frac{P_{kl}(c)}{P_{lk}(c)} = \frac{\prod\limits_{u \in \mathcal{U}} \left\{ 1 + \exp\left(\frac{r_k(u)h_l(u)}{c} \right) \right\}}{\prod\limits_{u \in \mathcal{U}} \left\{ 1 + \exp\left(\frac{r_l(u)h_k(u)}{c} \right) \right\}}$$

$$= \frac{\prod\limits_{u \in \mathcal{U}} \left\{ \exp\left(-\frac{r_k(u)h_l(u)}{2c} \right) + \exp\left(\frac{r_k(u)h_l(u)}{2c} \right) \right\} \prod\limits_{u \in \mathcal{U}} \exp\left(\frac{r_k(u)h_l(u)}{2c} \right)}{\prod\limits_{u \in \mathcal{U}} \left\{ \exp\left(-\frac{r_l(u)h_k(u)}{2c} \right) + \exp\left(\frac{r_l(u)h_k(u)}{2c} \right) \right\} \prod\limits_{u \in \mathcal{U}} \exp\left(\frac{r_l(u)h_k(u)}{2c} \right)}$$

$$= \frac{\prod\limits_{u \in \mathcal{U}} 2\cosh\left(\frac{r_k(u)h_l(u)}{2c} \right) \prod\limits_{u \in \mathcal{U}} \exp\left(\frac{r_k(u)h_l(u)}{2c} \right)}{\prod\limits_{u \in \mathcal{U}} 2\cosh\left(\frac{r_l(u)h_k(u)}{2c} \right) \prod\limits_{u \in \mathcal{U}} \exp\left(\frac{r_l(u)h_k(u)}{2c} \right)}. \tag{8.38}$$

Now, using the symmetry of the cosh, i.e. $\cosh(-x) = \cosh(x)$, and the following equality

$$\frac{\prod\limits_{u \in \mathcal{U}} \exp\left(\frac{r_k(u)h_l(u)}{2c} \right)}{\prod\limits_{u \in \mathcal{U}} \exp\left(\frac{r_l(u)h_k(u)}{2c} \right)} = \frac{\prod\limits_{u \in \mathcal{U}} \exp\left(-\frac{h_l(u)+2s_{\{u,u\}}l(u)}{2c} \right)}{\prod\limits_{u \in \mathcal{U}} \exp\left(-\frac{h_k(u)+2s_{\{u,u\}}k(u)}{2c} \right)},$$

which follows directly from the definitions of $h_i(u)$ and $r_i(u)$, $i \in \{k,l\}$, (8.38) reduces to

$$\frac{P_{kl}(c)}{P_{lk}(c)} = \frac{\prod\limits_{u \in \mathcal{U}} 2\cosh\left(\frac{h_l(u)}{2c} \right) \prod\limits_{u \in \mathcal{U}} \exp\left(-\frac{h_l(u)+2s_{\{u,u\}}l(u)}{2c} \right)}{\prod\limits_{u \in \mathcal{U}} 2\cosh\left(\frac{h_k(u)}{2c} \right) \prod\limits_{u \in \mathcal{U}} \exp\left(-\frac{h_k(u)+2s_{\{u,u\}}k(u)}{2c} \right)}.$$

Consequently, we have

$$q_k(c) \frac{P_{kl}(c)}{P_{lk}(c)} = \frac{1}{K_0(c)} \prod\limits_{u \in \mathcal{U}} 2\cosh\left(\frac{h_k(u)}{2c} \right) \exp\left(-\frac{h_k(u)+2s_{\{u,u\}}k(u)}{2c} \right)$$

$$\times \frac{\prod\limits_{u \in \mathcal{U}} 2\cosh\left(\frac{h_l(u)}{2c} \right) \prod\limits_{u \in \mathcal{U}} \exp\left(-\frac{h_l(u)+2s_{\{u,u\}}l(u)}{2c} \right)}{\prod\limits_{u \in \mathcal{U}} 2\cosh\left(\frac{h_k(u)}{2c} \right) \prod\limits_{u \in \mathcal{U}} \exp\left(-\frac{h_k(u)+2s_{\{u,u\}}k(u)}{2c} \right)}$$

$$= \frac{1}{K_0(c)} \prod_{u \in \mathcal{U}} 2 \cosh\left(\frac{h_l(u)}{2c}\right) \exp\left(-\frac{h_l(u) + 2s_{\{u,u\}}l(u)}{2c}\right)$$

$$= q_l(c),$$

which completes the proof of part (i).

To prove part (ii), the components of the stationary distribution given by (8.35) and (8.36) are rewritten as

$$q_k(c) = \frac{1}{K_0(c)} \exp\left(\sum_{u \in \mathcal{U}} -\frac{h_k(u) + 2s_{\{u,u\}}k(u) - \ln(2\cosh(h_k(u)))}{2c}\right).$$

Next, following the lines of Corollary 2.1 and using $\ln(\cosh(x)) \approx |x|$, for x small, it follows directly that

$$\lim_{c \downarrow 0} q_k(c) = \frac{1}{|\tilde{\mathcal{R}}_{opt}|} \chi_{(\tilde{\mathcal{R}}_{opt})}(k),$$

where $\tilde{\mathcal{R}}_{opt}$ denotes the set of configurations for which

$$-\sum_{u \in \mathcal{U}} \left\{ h_k(u) + 2s_{\{u,u\}}k(u) - |h_k(u)| \right\}$$

is maximal. This completes the proof of the theorem.   ∎

Apparently, a Boltzmann machine applying full synchronous parallelism does **not** maximize the consensus function but a different function, viz. the pseudo consensus function given in (8.37).

We mention that a similar property has been derived for the *Little model* [Little, 1974; Little & Shaw, 1978; Peretto, 1984], which is a spin glass like neural network model, based on a statistical physics approach. In the literature, only little attention has been paid to this special property. So far, its use is merely restricted to a few special cases in associative memories and for these cases it was shown that a similar behaviour occurs as in Hopfield's associative memory [Hopfield, 1982], which is comparable to the sequential Boltzmann machine; cf. [Little & Shaw, 1978; Peretto, 1984].

8.3.2 Asynchronous Parallelism

In an asynchronously parallel Boltzmann machine there is no global clocking scheme. Again, units may change their states simultaneously, but state transitions are no longer combined in successive cycles of the clocking scheme. It is again possible to distinguish between limited and unlimited parallelism.

Limited parallelism can be realized by introducing a *blocking scheme* which prohibits adjacent units to carry out state transitions simultaneously. Clearly, in this case, no erroneously calculated state transitions occur and it is intuitively evident that the Boltzmann machine will converge asymptotically to the set of optimal configurations, provided that all units have a finite probability of proposing state transitions. Then each configuration can be reached from every other configuration in a finite number of individual state transitions and no errors are made in computing the differences in consensus.

In the case of unlimited parallelism, no blocking scheme is used and evidently, erroneously calculated state transitions will occur. Clearly, as for unlimited synchronous parallelism, proving asymptotic convergence for a Boltzmann machine applying unlimited asynchronous parallelism is an open problem.

Due to the absence of synchronization, parallel state transitions cannot be modelled in sets of successive trials. In fact, individual state transitions can still be described as in the model of sequential Boltzmann machines, but the description of parallel state transitions, given in the previous section, no longer holds. In our opinion, the description of asynchronously parallel Boltzmann machines requires an approach that differs substantially from the Markov approach described in the previous sections. Within this context one might think of models describing asynchronous communication in concurrent or distributed processes.

A natural approach to describe this process is given by object oriented languages. The objects (in our case the computations carried out by individual units) are protected and only accessible to neighbouring objects, and they address each other in client/server relations [Dennis & Van Horn, 1966]. Even in the presence of erroneously calculated state transitions such an approach might work. For instance, Dolev, Lynch, Pinter, Stark & Weihl [1986] and Genesereth, Ginsberg & Rosenschein [1986] present general approaches to the description of distributed systems with faults. It is however not clear to what extent these approaches are useful for the description of asynchronously parallel Boltzmann machines.

Another approach is given by the more classical description of neural networks using concepts from statistical physics. This approach is based on the analogy between neural networks and spin glass models [Van Hemmen, 1982; Morgenstern, 1987; Toulouse, Dehaene & Changeux, 1986]. In such a *spin glass model* one considers an ensemble of spin particles, having spin 'up' or 'down', i.e. $S_i = \pm 1$. The corresponding *Hamiltonian* is given by

$$H = -\frac{1}{2}\sum_{i \leq j} J_{ij}S_iS_j + h_0 \sum_i S_i, \qquad (8.39)$$

where the J_{ij} denote the coupling strengths of the short-range spin-spin inter-action between two particles i and j, and h_0 denotes the strength of an external magnetic field.

The analogy with the Boltzmann machine is evident. The spins play the role of the states of the individual units and the Hamiltonian is the equivalent of the consensus function. The spin glass model described above is charac-teristic for systems studied in statistical physics. Hence, if the anology with neural networks can be pursued far enough, it permits the introduction into the theory of neural networks of such concepts as coupled differential equa-tions, phase transitions, correlation functions, mean field theories, etc., which provides a powerful tool for studying the continuous-time behaviour of neural networks; see for instance Peretto [1984]. This approach has been adopted by a number of researchers, and their work has shown interesting achievements. We mention studies related to the continuous-time behaviour of neural net-works by Hopfield & Tank [1985;1986] and Toulouse, Dehaene & Changeux [1986], and studies related to information storage capacities of neural net-works by Amit, Gutfreund & Sompolinsky [1987] and Personnaz, Guyon & Dreyfus[1985]. First applications of this approach to the description of Boltz-mann machines have been reported by Gutzmann [1987] and Levy & Adams [1987].

Clearly, both approaches discussed above are quite different. Basically they emphasize different aspects of neural networks, i.e. specification versus computation, and discrete versus continuous. In fact, both approaches in their present form fall short in presenting a full description of asynchronously par-allel Boltzmann machines, and additional research is needed to overcome this shortcoming.

However, in spite of this, asynchronous parallelism is generally consid-ered as the most promising approach to parallel Boltzmann machines since it yields a simple computational model for special-purpose VLSI implemen-tations, rendering global synchronization requirements superfluous. Future research has to reveal its significance.

Finally, we mention that the theoretical formulations presented in the re-mainder of this book are based on the sequential Boltzmann machine. The simulations, however, are based on the synchronously parallel Boltzmann ma-chine applying unlimited parallelism. For this we introduce a parallel cooling schedule in the next section.

8.3.3   A Parallel Cooling Schedule

To reduce global communication in a parallel implementation to a minimum, each unit is given its own cooling schedule. The parameters in the schedules are chosen such that they are constant or based on locally available information. Thus, no global communication is required to pass on the values of parameters during the consensus maximization. Furthermore, we assume unlimited parallelism, thus allowing for erroneously calculated state transitions. Applying limited parallelism would require a mechanism that prohibits adjacent units to carry out state transitions simultaneously. Such a unit-dependent or parallel cooling schedule can be applied to synchronously as well as asynchronously parallel Boltzmann machines.

The parallel cooling schedule presented below belongs to the class of 'conceptually simple' cooling schedules; see Section 4.1. It is based on a number of heuristic 'common sense' rules and does not follow a sophisticated theoretical concept. In our schedule, units independently generate a finite sequence of trials at descending values of a unit-dependent control parameter. Each unit u has its own sequence of values of the control parameter $(c_0^{(u)}, c_1^{(u)}, \ldots)$. The sequence is specified as follows.

- The start value $c_0^{(u)}$ is chosen as

$$c_0^{(u)} = \sum_{\{u,v\} \in C_u} |s_{\{u,v\}}|. \qquad (8.40)$$

   Using (8.10), it can be easily verified that, at this value of c, virtually all proposed state transitions are accepted, ensuring a sufficiently large start value of the control parameter; see also Section 4.1.

- The decrement function that determines a subsequent value of the control parameter from a current one ($c_{j+1}^{(u)}$ and $c_j^{(u)}$, respectively) is chosen as

$$c_{j+1}^{(u)} = \alpha c_j^{(u)}, \qquad (8.41)$$

   where α is a positive number smaller than but close to 1.

- The following stop criterion determines the final value of the control parameter. A unit stops generating trials if during K consecutive decrements of the control parameter no state transitions are accepted.

At each value of the control parameter, a unit generates L trials. Taking into account the time needed to propagate changes through the network, L is chosen proportional to the number of units in the Boltzmann machine.

The actual values of the parameters α, L, and K used in our computer simulations are given when we discuss the simulations; see Chapters 9, 10, and 11.

Clearly, as already pointed out in Chapter 4, the cooling schedule presented above can never guarantee convergence to the set of optimal configurations. In practice, however, it is found to result in high-quality configurations. With respect to the time complexity of the schedule we note that it is not possible to give a worst-case upper bound, due to the probabilistic stop criterion. Empirical analysis show that the average running time of a Boltzmann machine, applying the schedule given above, is usually given by a small power (≤ 2) in the number of units and connections in a Boltzmann machine; cf. Section 9.7.

8.4 A Taxonomy

As an introduction to Chapters 9, 10 and 11, we present in this section a taxonomy of applications of the Boltzmann machine. The taxonomy first distinguishes between fixed and variable connection strengths.

- *Fixed connection strengths.* The Boltzmann machine adjusts the states of the units to a given set of connection strengths that is fixed in time.

- *Variable connection strengths.* The Boltzmann machine adjusts the values of the connection strengths to the states of some the units which are fixed in time.

We associate two main problem areas with these two items, i.e. *search* and *learning*, respectively [Aarts & Korst, 1987].

We mention that Hinton [1986] distinguishes a third problem area called *representation.* An example of this is given by the work of Touretzky [1986], who developed a special-purpose Boltzmann machine, called BoltzCONS, for representing LISP *cons* structures. Here, we drop the subject of representation since it strongly focusses on programming aspects for artificial-intelligence purposes, which are considered beyond the scope of this book.

Search
Search can be divided into combinatorial optimization and classification.

In combinatorial optimization, one can define for each instance of a given optimization problem a Boltzmann machine for which each configuration corresponds one-to-one to a solution of the optimization problem. The data set that goes with the problem instance can be implemented by choosing an appropriate connection pattern and the appropriate connection strengths. Maximization of the consensus function thus becomes equivalent to solving the optimization problem at hand. We return to this in Chapter 9.

In classification the problem is to recognize an object as a member of a given subset, i.e. the object must be given a label denoting the subset to which it belongs. To that end, one can define a Boltzmann machine such that configurations uniquely represent objects and their classification labels. The subsets are represented by an appropriate connection pattern and an appropriate set of connection strengths. Maximization of the consensus function for a given object, which is presented to the Boltzmann machine by clamping the states of a subset of the units, results in classification of that object. We return to this in Chapter 10.

Learning

In a Boltzmann machine learning takes place by means of examples that are clamped into the states of a subset of the units, called the environmental units. The remaining units, the hidden units, are used to construct an internal representation that captures the regularities of the examples. For this, a learning algorithm is used that iteratively adjusts the connection strengths in a given Boltzmann machine until an appropriate internal representation is obtained. In this way a Boltzmann machine cannot only reproduce given examples (memory), it also exhibits associative capabilities. The structure of a Boltzmann machine can be chosen such that it is suitable for both supervised and unsupervised learning. The subject of learning is extensively discussed in Chapter 11.

In the following chapters several examples are given which illustrate the application of the Boltzmann machine in the areas mentioned above, i.e. combinatorial optimization, classification and learning, together with references to related work.

CHAPTER 9

Combinatorial Optimization and Boltzmann Machines

In this chapter we show how Boltzmann machines can be used to solve combinatorial optimization problems. The approach is based on the observation that the structure of many combinatorial optimization problems can be directly mapped onto the structure of a Boltzmann machine by choosing the right connection pattern and appropriate connection strengths. In this way maximizing the consensus in a Boltzmann machine is equivalent to finding optimal solutions of the corresponding optimization problem [Korst & Aarts, 1988].

If the inherent parallelism of a Boltzmann machine can be efficiently used for solving combinatorial optimization problems, then this approach can be viewed as a massively parallel implementation of simulated annealing. Simulations are carried out to compare Boltzmann machine implementations with sequential simulated annealing implementations for a number of different combinatorial optimization problems. The comparison is made for four problems, viz. the max cut, the independent set, the graph colouring, and the travelling salesman problem.

9.1 General Strategy

For our purposes we consider an instance of a combinatorial optimization problem as a tuple (S, S', f), where S denotes the finite set of solutions, $S' \subseteq S$ the set of feasible solutions, i.e. the set of solutions satisfying the constraints that go with the problem, and $f : S \rightarrow \mathbb{R}$ the cost function that assigns a real number to each solution; cf. Definition 5.3. Now, the problem is to find a feasible solution for which the cost function f is optimal, i.e. to find a solution

153

$i \in S'$ for which $f(i)$ is optimal.

A solution of a combinatorial optimization problem can be characterized by a finite set of discrete variables $X = \{x_1, \ldots, x_n\}$. If the domain of each variable $x_i \in X$ is given by a finite set X_i, then the set of solutions S is given by $S = X_1 \times X_2 \times \cdots \times X_n$.

To use the Boltzmann machine for solving combinatorial optimization problems, a bijective function $m : R \rightarrow S$ is defined which maps the set of configurations R onto the set of solutions S, by applying the following general strategy.

- Formulate the optimization problem as a 0-1 programming problem, i.e. formulate the problem such that $X_i = \{0, 1\}$, for $i = 1, \ldots, n$.

- Define a Boltzmann machine $B = (U, C)$, such that the state of each unit $u_i \in U = \{u_1, \ldots, u_n\}$ determines the value of variable x_i. If unit u_i is 'on' then $x_i = 1$; if u_i is 'off' then $x_i = 0$. Thus, $x_i = k(u_i)$. Clearly, this defines a bijective function m which maps the set of configurations R onto the set of solutions S.

- Define the set of connections C and the corresponding strengths such that the consensus function C is feasible and order-preserving.

The following definitions explain what we mean by a feasible and order-preserving consensus function.

Definition 9.1 Let $B = (U, C)$ be a Boltzmann machine that implements a combinatorial optimization problem (S, S', f) and let m be a bijective mapping from R onto S, then the consensus function C of the Boltzmann machine is called *feasible* if all local maxima of the consensus function C correspond to feasible solutions, i.e. if

$$m(\widehat{R}) \subseteq S', \tag{9.1}$$

where $\widehat{R}$ denotes the set of locally optimal configurations and

$$m(\widehat{R}) = \{i \in S \mid \exists k \in \widehat{R} : m(k) = i\}. \tag{9.2}$$

■

Feasibility of the consensus function implies that a feasible solution is always found, since a Boltzmann machine always converges to a configuration $k \in \widehat{R}$.

Definition 9.2 Let $B = (U, C)$ be a Boltzmann machine that implements a maximization problem (S, S', f) and let m be a bijective mapping from $\mathcal{R}$ onto S, then the consensus function C of the Boltzmann machine is called *order-preserving* if for all $k, l \in \mathcal{R}$, with $m(k), m(l) \in S'$, we have

$$f(m(k)) > f(m(l)) \Rightarrow C(k) > C(l). \tag{9.3}$$

Similarly, if the associated problem (S, S', f) is a minimization problem, the consensus function C is order-preserving if for all $k, l \in \mathcal{R}$, with $m(k), m(l) \in S'$, we have

$$f(m(k)) < f(m(l)) \Rightarrow C(k) > C(l). \tag{9.4}$$
∎

If the consensus function is feasible and order-preserving, then the consensus is maximal for configurations corresponding to an optimal (feasible) solution. Furthermore, under the same conditions, near-optimal local maxima of the consensus function correspond to near-optimal feasible solutions. This is very important, since only in the asymptotic case is a Boltzmann machine guaranteed to reach maximal consensus. Hence, these conditions are very important when using the Boltzmann machine for approximation purposes.

Many combinatorial optimization problems belong to the class of NP-complete problems. As all problems in the class of NP-complete problems are polynomially transformable into each other, it suffices, in theory, to show that it is possible to solve one NP-complete problem with a Boltzmann machine. We could then solve any NP-complete problem by transforming it to the problem for which a Boltzmann machine implementation is known. However, Boltzmann machine implementations tailored to a specific problem are usually more efficient. We therefore discuss Boltzmann machine implementations for a number of problems according to the general strategy discussed above. Four of these problems, viz. the max cut problem, the independent set problem, the graph colouring problem and the travelling salesman problem, have also been discussed in Chapter 5. In the following sections these problems are discussed in order of increasing (implementation) complexity.

9.2 The Max Cut Problem

Given a graph $G = (V, E)$ with positive weights on the edges, the max cut problem is defined as the problem of finding a partition of $V = \{1, 2, \ldots, n\}$ into disjoint sets V_0 and V_1 such that the sum of the weights of the edges that have one endpoint in V_0 and one endpoint in V_1 is maximal; cf. Definition 5.2.

To formulate the max cut problem as a 0-1 programming problem we need to define a number of variables. Let w_{ij} be the weight associated with the edge $\{i, j\}$ (by definition, $w_{ij} = w_{ji}$ and $w_{ij} = 0$ if $\{i, j\} \notin E$) and let x_i be a 0-1 variable defined as

$$x_i = \begin{cases} 1 & \text{if } i \in V_1 \\ 0 & \text{if } i \in V_0, \end{cases} \tag{9.5}$$

then the max cut problem can be formulated as

maximize

$$f(X) = \sum_{i=1}^{n} \sum_{j=i+1}^{n} w_{ij} \left\{ (1 - x_i)x_j + x_i(1 - x_j) \right\}. \tag{9.6}$$

Clearly, any solution corresponds uniquely to a partition of V into V_0 and V_1. Hence, the set of feasible solutions S' equals the total set of solutions S.

To implement the max cut problem on a Boltzmann machine we introduce a unit u_i for each variable x_i. The set of connections is taken as the union of two disjoint sets, viz.

- *bias connections*

$$C_b = \{\{u_i, u_i\} \mid i \in V\}, \quad \text{and} \tag{9.7}$$

- *weight connections*

$$C_w = \{\{u_i, u_j\} \mid \{i, j\} \in E\}. \tag{9.8}$$

Next, the connection strengths are chosen such that maximizing the consensus function corresponds to maximizing the cost function of (9.6). Clearly, as $S' = S$, the consensus function C is feasible by definition. Theorem 9.1 yields values for the connection strengths that make the consensus function order-preserving.

Theorem 9.1 *Let b_i be the sum of the weights of all edges incident with vertex i, i.e. $b_i = \sum_{j=1}^{n} w_{ij}$, and let*

$$\forall \{u_i, u_i\} \in C_b \; : \; s_{\{u_i, u_i\}} = \; b_i, \quad \text{and} \tag{9.9}$$
$$\forall \{u_i, u_j\} \in C_w \; : \; s_{\{u_i, u_j\}} = -2w_{ij}, \tag{9.10}$$

then the consensus function C is order-preserving.

Proof To prove this theorem we show that, for this particular choice of the connection strengths, the consensus function equals the cost function of (9.6). If the connection strengths are chosen as indicated above, then the consensus function can be written as

$$C(k) = \sum_{\{u_i,u_i\}\in C_b} b_i\, k^2(u_i) + \sum_{\{u_i,u_j\}\in C_w} -2\, w_{ij}\, k(u_i)\, k(u_j),$$

which can be rewritten into

$$C(k) = \sum_{i=1}^{n}\sum_{j=1}^{n} w_{ij}x_i + \sum_{i=1}^{n}\sum_{j=i+1}^{n} -2w_{ij}x_i x_j.$$

Since, by definition, $w_{ij} = w_{ji}$ for all i, j, and $w_{ii} = 0$ for all i, we have

$$C(k) = \sum_{i=1}^{n}\sum_{j=i+1}^{n} w_{ij}(x_i + x_j) + \sum_{i=1}^{n}\sum_{j=i+1}^{n} -2w_{ij}x_i x_j,$$

which is identical to the cost function of (9.6). ∎

An example of a max cut problem and the corresponding Boltzmann machine is given in Figure 9.1.

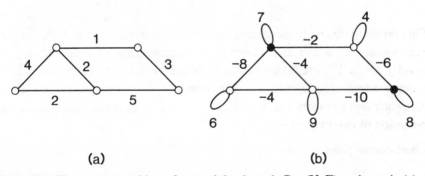

(a)   (b)

Figure 9.1: The max cut problem: for a weighted graph $G = (V, E)$ as shown in (a), a Boltzmann machine is constructed whose structure is isomorphic to G (b). Consensus maximization gives the maximal cut, indicated by the units that are 'on'. These units are drawn as solid dots.

9.3 The Independent Set Problem

Given a graph $G = (V, E)$, then the independent set problem is to find an independent set of maximal size, i.e. a subset $V' \subseteq V$, such that for all $i, j \in V'$ the edge $\{i, j\}$ is not in E and such that $|V'|$ is maximal, cf. Definition 5.4.

In a 0-1 programming formulation the problem can be written as follows. Let the 0-1 variables x_i be defined by

$$x_i = \begin{cases} 1 & \text{if } i \in V' \\ 0 & \text{if } i \notin V', \end{cases} \qquad (9.11)$$

then the independent set problem can be written as:

maximize

$$f(X) = \sum_{i=1}^{n} x_i, \qquad (9.12)$$

subject to

$$x_i x_j e_{ij} = 0, \quad i, j = 1, \ldots, n, \qquad (9.13)$$

where n denotes the number of vertices and e_{ij} is defined by

$$e_{ij} = \begin{cases} 1 & \text{if } \{i, j\} \in E \\ 0 & \text{if } \{i, j\} \notin E. \end{cases} \qquad (9.14)$$

This problem can be implemented on a Boltzmann machine as follows. For each variable x_i a unit u_i is defined. To guarantee that no adjacent vertices i and j are in V', inhibitory connections are introduced which connect the corresponding units u_i and u_j. Furthermore, bias connections are used to give each unit a tendency to be 'on'. Hence, the set of connections is taken as the union of two disjoint sets, viz.

- *bias connections*

$$C_b = \{\{u_i, u_i\} \mid i \in V\}, \quad \text{and} \qquad (9.15)$$

- *inhibitory connections*

$$C_h = \{\{u_i, u_j\} \mid \{i, j\} \in E\}. \qquad (9.16)$$

Next, the connection strengths are chosen such that maximizing the consensus function corresponds to maximizing the cost function of (9.12) subject to the constraints of (9.13). Theorem 9.2 yields values for the connection strengths for which the consensus function is feasible and order-preserving.

Theorem 9.2 *Let β be a positive number and let*

$$\forall \{u_i, u_i\} \in C_b \; : \quad s_{\{u_i, u_i\}} = \beta, \quad and \tag{9.17}$$

$$\forall \{u_i, u_j\} \in C_h \; : \quad s_{\{u_i, u_j\}} < -\beta, \tag{9.18}$$

then the consensus function C is feasible and order-preserving.

Proof To prove this theorem we distinguish between two disjoint subsets of the configuration space of the Boltzmann machine, i.e. $\mathcal{R} = \mathcal{R}_A \cup \mathcal{R}_B$ with $\mathcal{R}_A \cap \mathcal{R}_B = \emptyset$, where $\mathcal{R}_A$ and $\mathcal{R}_B$ denote the sets of configurations corresponding to feasible and infeasible solutions, respectively. Now, it can be straightforwardly proved that C is feasible by showing that

$$\forall k \in \mathcal{R}_B \; \exists u_i \in \mathcal{U} \; : \quad \Delta C_k(u_i) > 0.$$

Indeed, if $k \in \mathcal{R}_B$, then an inhibitory connection $\{u_i, u_j\}$ is activated. Changing the state of one of the units u_i and u_j, say u_i, changes the consensus by $\Delta C_k(u_i) = C(k_{u_i}) - C(k) \geq -s_{\{u_i, u_i\}} - s_{\{u_i, u_j\}} > 0$.

The consensus function C is order-preserving. This follows directly from the fact that for any configuration $k \in \mathcal{R}_A$, i.e. for any feasible solution, the consensus function can be written as

$$
\begin{aligned}
C(k) \;&=\; \sum_{\{u_i, u_i\} \in C_b} s_{\{u_i, u_i\}} \, k^2(u_i) \\
&=\; \sum_{u_i \in \mathcal{U}} \beta \, k(u_i) \\
&=\; \sum_{i \in V'} \beta \, x_i,
\end{aligned}
$$

which scales linear with the cost function of (9.12). Clearly, the consensus function and the cost function are identical if β is chosen equal to 1. ∎

Figure 9.2 illustrates the implementation of the independent set problem on a Boltzmann machine. Equivalent problems, such as the vertex cover problem and the clique problem, can be solved on a Boltzmann machine in a similar way, as they can easily be transformed to the independent set problem [Garey & Johnson, 1979].

9.4 The Graph Colouring Problem

Given a graph $G = (V, E)$, then the graph colouring problem is to find a minimal colouring, i.e. a mapping $g : V \rightarrow \{1, 2, \ldots, l\}$ such that $g(i) \neq g(j)$ for all $i, j \in V$ for which $\{i, j\} \in E$, and l is minimal; cf. Definition 5.5.

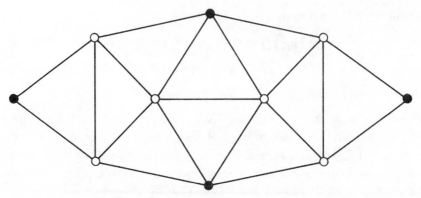

Figure 9.2: The independent set problem: for a graph $G = (V, E)$ a Boltzmann machine is constructed such that it is isomorphic to G. After maximizing the consensus, using an appropriate choice of connection strengths, the units that are 'on' (drawn as solid dots) make up the maximal independent set.

To implement this problem on a Boltzmann machine we use Theorem 5.1 which states that for an appropriate choice of the weights in the set $W = \{w_1, \ldots, w_{\Delta+1}\}$, associated with $\Delta + 1$ different colours, where Δ denotes the maximum degree of G, the graph colouring problem is equivalent to the following problem:

maximize

$$f(\mathcal{X}) = \sum_{l=1}^{\Delta+1} \sum_{i=1}^{n} w_l x_{il}, \qquad (9.19)$$

subject to

$$x_{il} x_{jl} e_{ij} = 0, \quad i, j = 1, \ldots, n; l = 1, \ldots, \Delta + 1, \quad \text{and} \qquad (9.20)$$

$$\sum_{l=1}^{\Delta+1} x_{il} = 1, \quad i = 1, \ldots, n, \qquad (9.21)$$

where n denotes the number of vertices, w_l denotes the weight associated with colour l, and where x_{il} is the 0-1 variable indicating whether or not vertex i is given colour l and e_{ij} denotes whether or not vertices i and j are adjacent. Now, the graph colouring problem is implemented on a Boltzmann machine in the following way. A Boltzmann machine is chosen consisting of $\Delta + 1$ (horizontal) layers, each layer corresponding to a specific colour. The number

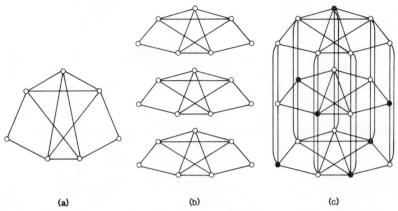

Figure 9.3: The graph colouring problem: for a graph $G = (V, E)$ as shown in (a), a Boltzmann machine is constructed consisting of $\Delta+1$ layers. Each layer is isomorphic to G and corresponds to a specific colour (b). All units corresponding to the same $i \in V$ are mutually connected (c). For reasons of clarity the number of layers is chosen here to be smaller than $\Delta+1$. The units that are drawn as solid dots correspond to a minimal colouring.

of units in each layer is equal to the number of vertices in the original graph. Thus, for each vertex i and each colour l a unit u_{il} is defined. The state of unit u_{il} determines the value of variable x_{il}. To guarantee that the same colour is not assigned to adjacent vertices, (horizontal) inhibitory connections are defined to connect units within the same layer. To guarantee that no more than one colour is assigned to a vertex, all units that correspond to the same vertex are mutually connected by (vertical) inhibitory connections. The latter run between the different layers. An example illustrating the choice of units and connections is given in Figure 9.3. Thus, the set of connections is taken as the union of three disjoint sets, viz.

- *bias connections*

$$C_b = \{\{u_{il}, u_{il}\} \mid i \in V\}, \tag{9.22}$$

- *horizontal inhibitory connections*

$$C_h = \{\{u_{il}, u_{jl}\} \mid \{i, j\} \in E\}, \quad \text{and} \tag{9.23}$$

- *vertical inhibitory connections*

$$C_v = \{\{u_{il}, u_{im}\} \mid i \in V \wedge l \neq m\}, \tag{9.24}$$

where $l, m \in \{1, \ldots, \Delta + 1\}$.

Next, the connection strengths are chosen such that maximizing the consensus function corresponds to maximizing the cost function of (9.19) subject to the constraints of (9.20) and (9.21). A possible choice of the connection strengths is given by Theorem 9.3.

Theorem 9.3 *Let the weights w_j be positive real numbers and chosen such that the conditions of Theorem 5.1 are satisfied and let*

$$\forall \{u_{il}, u_{il}\} \in C_b \quad : \quad s_{\{u_{il}, u_{il}\}} = w_l, \tag{9.25}$$

$$\forall \{u_{il}, u_{jl}\} \in C_h \quad : \quad s_{\{u_{il}, u_{jl}\}} < -w_l, \quad and \tag{9.26}$$

$$\forall \{u_{il}, u_{im}\} \in C_v \quad : \quad s_{\{u_{il}, u_{im}\}} < -\min\{w_l, w_m\}, \tag{9.27}$$

then the consensus function C is feasible and order-preserving.

Proof To prove the theorem we distinguish between four subsets of the configuration space of the Boltzmann machine, i.e. $R = R_A \cup R_B \cup R_C \cup R_D$, with $R_A \cap (R_B \cup R_C \cup R_D) = \emptyset$, where the subsets are defined as follows:

(1) R_A denotes the set of configurations corresponding to feasible solutions,

(2) $R_B = \{k \in R \mid \exists i, j, l : k(u_{il}) = 1 \land k(u_{jl}) = 1 \land e_{ij} = 1\}$ denotes the set of configurations in which at least two adjacent vertices have the same colour,

(3) $R_C = \{k \in R \mid \exists i, l, m : k(u_{il}) = 1 \land k(u_{im}) = 1 \land l \neq m\}$ denotes the set of configurations in which at least one vertex has two or more colours, and

(4) $R_D = \left\{k \in R \mid \exists i : \sum_{l=1}^{\Delta+1} k(u_{il}) = 0\right\}$ denotes the set of configurations in which at least one vertex does not have a colour.

It can be proved that the consensus function C is feasible by showing that

$$\forall k \in (R_B \cup R_C \cup R_D) \; \exists u \in \mathcal{U} \; : \; \Delta C_k(u) > 0.$$

To prove this we distinguish between the following three situations.

- If $k \in R_B$ then $\exists i, j, l : k(u_{il}) = 1 \land k(u_{jl}) = 1 \land e_{ij} = 1$.
  Clearly, in such a configuration k, the horizontal inhibitory connection $\{u_{il}, u_{jl}\}$ is activated. From the definition of the strength of a horizontal inhibitory connection it is apparent that changing the state of one of the units u_{il} or u_{jl} increases the consensus.

- If $k \in \mathcal{R}_C$ then $\exists i, l, m : k(u_{il}) = 1 \wedge k(u_{im}) = 1 \wedge l \neq m$.

  Clearly, in such a configuration k, the vertical inhibitory connection $\{u_{il}, u_{im}\}$ is activated. The strength of a vertical inhibitory connection is chosen such that changing the state of unit u_{il} (assuming that $w_l < w_m$) increases the consensus.

- If $k \in \mathcal{R}_D \setminus \mathcal{R}_C$ then $\exists i : \sum_{l=1}^{\Delta+1} u_{il} = 0$.

  Clearly, in such a configuration k, no colour is assigned to vertex i. Given that the number of layers is chosen equal to $\Delta + 1$ and given that $k \notin \mathcal{R}_C$, then there must exist a layer l in which unit u_{il} can be 'turned on' without activating any inhibitory connections. Clearly, this increases the consensus.

Furthermore, the consensus function C is order-preserving. This follows immediately from the fact that for any configuration $k \in \mathcal{R}_A$, i.e. for any feasible solution, the consensus function can be written as

$$
\begin{aligned}
C(k) &= \sum_{\{u_{il}, u_{il}\} \in C_b} s_{\{u_{il}, u_{il}\}}\, k^2(u_{il}) \\
&= \sum_{l=1}^{\Delta+1} \sum_{i=1}^{n} w_l\, x_{il},
\end{aligned}
$$

which is identical to the cost function of (9.19). ∎

9.5 The Clique Partitioning and Clique Covering Problems

The clique partitioning and clique covering problems are closely related to the graph colouring problem [Garey & Johnson, 1979; Kou, Stockmeyer & Wong, 1978] and can be treated accordingly. Here, we restrict ourselves to presenting their problem formulations.

Definition 9.3 (Clique partitioning problem) Given a graph $G = (V, E)$, the clique partitioning problem is to find a minimal partition into cliques, i.e. a partition of V into a minimal number of subsets $V_1, V_2, \ldots, V_k$ such that each V_i induces a complete subgraph or clique of G. ∎

Definition 9.4 (Clique covering problem) Given a graph $G = (V, E)$, the clique covering problem is to find a minimal clique cover, i.e. a minimal number of subsets $V_1, V_2, \ldots, V_k$ of V such that each V_k induces a complete subgraph of G and such that for each $\{i, j\} \in E$ there is some V_k that contains both i and j.   ■

Figures 9.4 and 9.5 illustrate a possible implementation of these problems on a Boltzmann machine via transformation to the graph colouring problem.

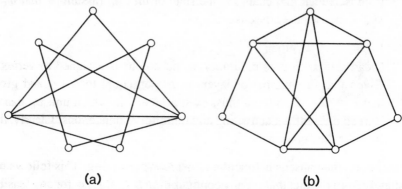

(a)   (b)

Figure 9.4: The clique partitioning problem: for a graph $G = (V, E)$ as shown in (a), a Boltzmann machine is chosen reflecting the structure of the complementary graph $G^c = (V, E^c)$ as shown in (b), where $E^c = \{\{i, j\} \mid i, j \in V \wedge \{i, j\} \notin E\}$. Colouring G^c with a minimal number of colours directly corresponds to partitioning G into a minimal number of cliques. Note that the graph in (b) equals the graph used in Figure 9.3.

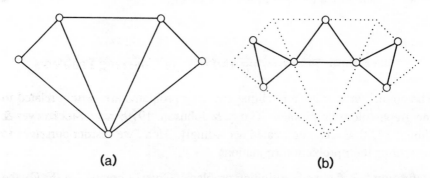

(a)   (b)

Figure 9.5: The clique covering problem: for a graph $G = (V, E)$ as shown in (a), a clique cover is equivalent to a clique partition of the graph G' shown in (b), which is defined as follows. Each vertex in G' corresponds to an edge in G and two vertices in G' are joined by an edge if the endpoints of the corresponding edges in G form a clique. A minimal clique partition of G' then yields a minimal clique cover of G. Note that G' is isomorphic to the graph used in Figure 9.4a.

Figure 9.4 illustrates the relation between the clique partitioning problem and the graph colouring problem, while Figure 9.5 illustrates the relation between the clique covering problem and the clique partitioning problem.

9.6 The Travelling Salesman Problem

Let n be the number of cities and $D = [d_{ij}]$ the distance matrix whose entry d_{ij} denotes the distance between city i and city j. Then the travelling salesman problem is to find a tour of minimal length visiting each of the n cities exactly once; cf. Example 1.1 and Definition 5.1.

Here, we formulate the travelling salesman problem as a 0-1 programming problem, by defining it as a quadratic assignment problem [Garfinkel, 1985]. An alternative approach for implementing the travelling salesman problem on a Boltzmann machine, by defining it as a linear assignment problem, is given by Aarts & Korst [1988a].

Let x_{ip} be a 0-1 variable defined as

$$x_{ip} = \begin{cases} 1 & \text{if the tour visits city } i \text{ at the } p^{th} \text{ position} \\ 0 & \text{otherwise,} \end{cases} \tag{9.28}$$

then the travelling salesman problem can be formulated as a quadratic assignment problem in the following way:

minimize

$$f(X) = \sum_{i,j,p,q=0}^{n-1} a_{ijpq}\, x_{ip}\, x_{jq}, \tag{9.29}$$

subject to

$$\sum_{i=0}^{n-1} x_{ip} = 1, \qquad p = 0,\dots,n-1, \quad \text{and} \tag{9.30}$$

$$\sum_{p=0}^{n-1} x_{ip} = 1, \qquad i = 0,\dots,n-1, \tag{9.31}$$

where a_{ijpq} is defined as

$$a_{ijpq} = \begin{cases} d_{ij} & \text{if } q = (p+1) \bmod n \\ 0 & \text{otherwise.} \end{cases} \tag{9.32}$$

To implement the travelling salesman problem on a Boltzmann machine we introduce for each variable x_{ip} a unit u_{ip}. Again, the set of connections C is chosen as a union of three disjoint sets, viz.

- *bias connections*

$$C_b = \{\{u_{ip}, u_{ip}\}\},\qquad(9.33)$$

- *distance connections*

$$C_d = \{\{u_{ip}, u_{jq}\} \mid i \neq j \wedge q = (p+1) \bmod n\},\quad \text{and}\qquad(9.34)$$

- *inhibitory connections*

$$C_h = \{\{u_{ip}, u_{jq}\} \mid (i = j \wedge p \neq q) \vee (i \neq j \wedge p = q)\},\qquad(9.35)$$

where $i, j, p, q = 0, \ldots, n - 1$. Consequently, the total number of connections equals $2n^3 - n^2$ (for $n > 2$). Next, the connection strengths are chosen such that maximizing the consensus function C corresponds to maximizing the cost function of (9.29) subject to the constraints of (9.30) and (9.31). Theorem 9.4 states that the strengths of the connections can be chosen such that the consensus function C is feasible and order-preserving.

Theorem 9.4 *Let the connection strengths be chosen such that*

$$\forall\{u_{ip}, u_{ip}\} \in C_b \quad : \quad s_{\{u_{ip}, u_{ip}\}} > \max\{d_{ik} + d_{il} \mid k \neq l\},\qquad(9.36)$$

$$\forall\{u_{ip}, u_{jq}\} \in C_d \quad : \quad s_{\{u_{ip}, u_{jq}\}} = -d_{ij}, \; and\qquad(9.37)$$

$$\forall\{u_{ip}, u_{jq}\} \in C_h \quad : \quad s_{\{u_{ip}, u_{jq}\}} < -\min\{s_{\{u_{ip}, u_{ip}\}}, s_{\{u_{jq}, u_{jq}\}}\},\quad(9.38)$$

then the consensus function C is feasible and order-preserving.

Proof To prove the theorem we distinguish between three subsets of the configuration space R of the Boltzmann machine, i.e. $R = R_A \cup R_B \cup R_C$, where

(1) R_A denotes the set of configurations corresponding to feasible solutions,

(2) R_B denotes the set of configurations corresponding to solutions for which a city is visited at least twice or for which two or more cities are visited at the same time, i.e. if $k \in R_B$, then

$$\exists i : \sum_{p=0}^{n-1} k(u_{ip}) \geq 2, \quad \text{or}$$

$$\exists p : \sum_{i=0}^{n-1} k(u_{ip}) \geq 2, \quad \text{and}$$

(3) $\mathcal{R}_C$ denotes the set of configurations corresponding to solutions for which a city is not visited, i.e. if $k \in \mathcal{R}_C$, then

$$\exists i : \sum_{p=0}^{n-1} k(u_{ip}) = 0.$$

From the definitions given above it follows directly that

$$\mathcal{R}_A \cap (\mathcal{R}_B \cup \mathcal{R}_C) = \emptyset.$$

We now prove the feasibility of the consensus function C by showing that

$$\forall k \in \mathcal{R}_B \cup \mathcal{R}_C \ \exists u \in \mathcal{U} : \ \Delta C_k(u) > 0,$$

thus implying that all local maxima of the consensus function C correspond to feasible solutions. The proof runs as follows.

- If $k \in \mathcal{R}_B$, then there exists an inhibitory connection $\{u_{ip}, u_{jq}\}$ which is activated. Clearly, if the connection strength of $\{u_{ip}, u_{jq}\}$ is chosen as indicated by the theorem, then changing the state of one of the units u_{ip} and u_{jq} (whichever has the lowest bias), raises the consensus.

- If $k \in \mathcal{R}_C \setminus \mathcal{R}_B$, then there exist a city i and a position p for which

$$\sum_{j=0}^{n-1} k(u_{jp}) = 0 \land \sum_{q=0}^{n-1} k(u_{iq}) = 0.$$

If the strengths of the bias and distance connections are chosen as indicated, then changing the state of unit u_{ip} raises the consensus. Changing the state of unit u_{ip} only activates the bias connection $\{u_{ip}, u_{ip}\}$ and at most two distance connections.

It is easy to prove that the consensus function C is order-preserving. This follows directly from the fact that for any configuration $k \in \mathcal{R}_A$ the consensus function can be written as

$$C(k) = \sum_{\{u_{ip}, u_{ip}\} \in \mathcal{C}_b} s_{\{u_{ip}, u_{ip}\}} \, k^2(u_{ip})$$

$$+ \sum_{\{u_{ip}, u_{jq}\} \in \mathcal{C}_d} s_{\{u_{ip}, u_{jq}\}} \, k(u_{ip}) \, k(u_{jq}).$$

As the first term is identical for all $k \in \mathcal{R}_A$, the expression can be rewritten as

$$
\begin{aligned}
C(k) &= K + \sum_{\{u_{ip}, u_{jq}\} \in C_d} s_{\{u_{ip}, u_{jq}\}}\, k(u_{ip})\, k(u_{jq}) \\
&= K - \sum_{\{u_{ip}, u_{jq}\} \in C_d} d_{ij}\, k(u_{ip})\, k(u_{jq}),
\end{aligned}
$$

for some $K > \sum_{i=0}^{n-1} \max\{d_{ik} + d_{il} \mid k \neq l\}$. The second term of this expression is equivalent to the cost function of (9.29), i.e. the length of the corresponding tour. Consequently, tours of shorter length correspond to configurations $k \in \mathcal{R}_A$ with a higher consensus. Moreover, an optimal tour corresponds to a configuration with maximal consensus.   ∎

We end this section with some remarks. For most feasible solutions, the value of the constant K is much larger than the length of the tour that corresponds to such a solution. Thus, the differences in consensus between individual tours are relatively small, which results in a slow convergence of the optimization process to near-optimal tours. However, a speed-up of the convergence can be obtained by a minor modification of the connection strengths of the bias connections. Instead of choosing the strength of a bias connection $\{u_{ip}, u_{ip}\}$ greater than $\max\{d_{ki} + d_{il} \mid k \neq l\}$, it is chosen equal to the average value of the sum of the distances, i.e.

$$
s_{\{u_{ip}, u_{ip}\}} = \sum_{k=0, k \neq i}^{n-1} \sum_{l=k+1, l \neq i}^{n-1} \frac{2\,(d_{ki} + d_{il})}{(n^2 - 3n + 2)}. \tag{9.39}
$$

For this choice, the feasibility of the consensus function C can no longer be guaranteed. However, the feasibility of C still holds for configurations in the near-optimal region, provided the positioning of the cities is sufficiently random. Since this is the regime of interest, this does not significantly limit the usefulness of the choice of connection strengths. The computer simulations presented in Section 9.7.2 are based on this choice and, indeed, we have observed that near-optimal results can be obtained within a computation time that is substantially shorter than the computation time required by simulations if the connection strengths are given by Theorem 9.4. In some cases the final result obtained by a simulation does not correspond to a tour, since local maxima of the consensus function are no longer restricted to $\mathcal{R}_A$. It is, however, observed that these situations do not occur frequently.

9.7 Numerical Results

With respect to combinatorial optimization the Boltzmann machine can be viewed as a massively parallel implementation of simulated annealing. To demonstrate the practical feasibility of the model of Boltzmann machines for approximating optimal solutions of combinatorial optimization problems, we simulated the Boltzmann machine for four of the combinatorial optimization problems discussed above, viz. the max cut problem, the independent set problem, the graph colouring problem and the travelling salesman problem. The first three problems have in common that they use graphs as problem instances, and therefore the numerical results concerning these problems are jointly discussed in Section 9.7.1. The numerical results for the travelling salesman problem are discussed separately in Section 9.7.2.

9.7.1 Graph Problems

The results obtained by the Boltzmann machine for the three graph problems discussed in this section are compared with results obtained by sequential implementations of the simulated annealing algorithm as discussed in Sections 5.1.2, 5.1.3, and 5.1.4. Both the Boltzmann machine and simulated annealing are implemented in PASCAL on a VAX 11/785.

The problem instances are randomly generated graphs $G = (V, E)$, with a fixed set of vertices V ($|V| = 50, 100, 150, 200, 250$). The edges are chosen independently with a probability p of being in E. The probability p is chosen equal to $\frac{10}{|V|-1}$, such that the expected degree for all vertices is equal to 10. In this way, the average connectivity is the same for all problem instances. Thus, we concentrate on investigating the time complexities of both algorithms as a function of the number of vertices only. The dependence of the performance on the average connectivity is not considered here, since it strongly depends on the machine architecture of the parallel computer on which the model is eventually emulated. For each number of vertices, five different problem instances are randomly generated, resulting in 25 different problem instances. For the max cut problem edges are given an integer weight randomly chosen from $\{1, \ldots, 10\}$.

In the Boltzmann machine parallelism is simulated according to the unlimited synchronous approach; see Section 8.3.1. During each trial of the optimization process, a subset of units $\mathcal{U}'$ is randomly chosen, containing a fixed fraction q of the total set of units. For each unit $u \in \mathcal{U}'$ the corresponding $\Delta C_k(u)$ is calculated, based on the configuration k, the outcome of the previous trial. Next, the states of the units in $\mathcal{U}'$ are adjusted according to the acceptance probability of (8.10), resulting in a configuration l as the outcome of the present trial. The fraction q is chosen equal to $\frac{2}{3}$.

The parameters of the cooling schedule used in the sequential simulated annealing algorithm are chosen such that high-quality solutions are obtained. We mention that iterative improvement algorithms, using the same generation mechanisms, find solutions for the problem instances at hand, deviating on the average more than 20% in cost from the solutions obtained by simulated annealing. The parameters of the cooling schedule used in the sequential simulated annealing algorithm are chosen as follows (see Section 4.2): $\chi_0 = 0.9$, $\delta = 0.1$ and $\varepsilon_s = 10^{-5}$.

The parameters of the cooling schedule used in the Boltzmann machine are chosen such that the final results obtained by the Boltzmann machine are comparable to the results obtained by simulated annealing. In this way the discussion on the performance can be concentrated solely on the computation times. For both implementations the computation times are proportional to the number of executed trials. For the Boltzmann machine we use the parallel cooling schedule described in Section 8.3.3 with the following parameters: $\alpha = 0.95$, $L = \frac{1}{4}|V|$, and $K = 10$. The simulated annealing implementations are as indicated in Sections 5.1.2, 5.1.3, and 5.1.4.

For each problem instance both the Boltzmann machine and simulated annealing are run 10 times, each time using a different initial configuration. Average final values of the consensus C of the Boltzmann machine and the cost function f of simulated annealing together with the corresponding standard deviations, σ_C and σ_f, are given in Tables 9.1, 9.2 and 9.3 for the various problems. The values of the consensus and the cost function can be directly compared, as both algorithms always produced feasible solutions. In that case, the consensus C and the cost function f are identical. The tables furthermore give the average computation times $\bar{t}$ and the corresponding standard deviations σ_t. The compution times of the Boltzmann machine algorithm are obtained by dividing the computation times of the sequential implementation by the number of units that operate in parallel. The times in the tables are given in seconds.

From the tables the following conclusions can be drawn. First, we observe that the Boltzmann machine is able to obtain results that are comparable, in quality, to the results obtained by simulated annealing. This is in itself not trivial. For the travelling salesman problem, for instance, we see in the next section, that it is impossible to obtain comparable results in reasonable amounts of computation time.

Problem	Simulated Annealing				Boltzmann Machine			
instance	$\overline{f}$	σ_f	$\overline{t}$	σ_t	$\overline{C}$	σ_C	$\overline{t}$	σ_t
N50N1	985.6	13.0	3.1	0.5	990.4	7.3	0.2	0.06
N50N2	924.4	15.7	2.7	0.3	926.5	8.1	0.1	0.04
N50N3	932.4	6.6	2.7	0.2	930.1	8.0	0.2	0.05
N50N4	1022.8	10.1	3.3	0.7	1032.4	3.0	0.1	0.06
N50N5	1023.2	9.8	3.2	0.3	1028.3	2.5	0.1	0.06
N100N1	1901.1	9.2	11.0	1.4	1912.2	7.1	0.4	0.10
N100N2	1856.8	17.5	11.0	1.6	1865.0	13.0	0.4	0.05
N100N3	1861.7	8.6	9.7	1.0	1863.5	8.3	0.3	0.07
N100N4	1922.5	11.0	10.7	1.0	1918.8	13.0	0.4	0.11
N100N5	1799.5	13.9	10.4	1.9	1806.7	10.2	0.3	0.08
N150N1	2837.5	8.8	21.3	3.6	2844.7	12.3	0.6	0.18
N150N2	2996.6	15.0	24.5	4.8	2996.2	20.9	0.6	0.08
N150N3	3105.3	15.1	27.1	5.2	3113.1	18.7	0.6	0.15
N150N4	2918.0	18.0	23.3	4.7	2906.1	17.8	0.7	0.07
N150N5	2943.4	20.4	19.7	2.8	2957.5	26.9	0.7	0.14
N200N1	3730.0	23.4	36.4	4.7	3731.2	24.0	0.7	0.14
N200N2	3974.8	18.7	41.6	7.0	3964.8	26.9	1.0	0.15
N200N3	4118.0	4.2	35.1	3.4	4114.3	25.3	0.7	0.12
N200N4	3811.6	14.8	36.6	5.2	3817.7	18.3	0.9	0.11
N200N5	3925.1	16.9	41.8	6.4	3927.5	14.2	0.9	0.13
N250N1	5183.0	26.5	67.9	12.0	5195.0	14.6	1.3	0.22
N250N2	5343.6	13.6	63.4	6.5	5335.9	29.5	1.4	0.12
N250N3	5074.2	13.3	59.0	10.0	5068.4	21.6	1.3	0.12
N250N4	5338.4	22.6	64.4	17.6	5345.6	21.7	1.2	0.28
N250N5	5078.4	15.0	63.2	12.3	5096.0	11.2	1.2	0.15

Table 9.1: Numerical results for the max cut problem.

To give an indication of the quality of the final results obtained by both algorithms, we mention that for the independent set problem the algorithms were able to find independent sets which are on average three times as large as the expected average independent set. The expected average size of an independent set for this type of random graph equals $\frac{|V|}{10}$.

Furthermore, we observe that, for the problem instances we investigated, the Boltzmann machine yields a speed-up over the sequential simulated annealing algorithm ranging from 20 up to 400.

From the tables the average-case time complexities can be estimated. For the max cut problem, the average-case time complexities of the simulated annealing algorithm and the Boltzmann machine are estimated to be $O(|V|^{2.0})$

Problem	Simulated Annealing				Boltzmann Machine			
instance	$\bar{f}$	σ_f	$\bar{t}$	σ_t	$\overline{C}$	σ_C	$\bar{t}$	σ_t
N50N1	13.0	0.0	3.4	0.3	12.9	0.3	0.1	0.02
N50N2	13.9	0.3	3.2	0.2	13.9	0.3	0.1	0.02
N50N3	15.0	0.0	3.3	0.2	14.9	0.3	0.1	0.02
N50N4	13.6	0.7	3.4	0.2	13.8	0.6	0.1	0.02
N50N5	12.9	0.3	3.6	0.2	13.0	0.0	0.1	0.02
N100N1	31.3	0.4	11.2	0.9	31.4	0.5	0.2	0.03
N100N2	30.8	0.7	10.7	0.5	31.3	0.8	0.1	0.03
N100N3	31.3	0.5	10.5	0.8	31.7	0.5	0.2	0.03
N100N4	29.2	1.0	10.6	0.5	28.7	0.6	0.1	0.04
N100N5	29.7	0.5	10.1	0.3	29.9	0.3	0.2	0.03
N150N1	44.6	0.5	21.1	1.4	44.8	0.7	0.3	0.03
N150N2	45.7	0.6	22.2	1.5	45.5	0.5	0.3	0.03
N150N3	41.3	0.8	22.1	1.2	41.8	0.4	0.3	0.02
N150N4	45.0	0.8	21.5	1.1	45.2	0.6	0.2	0.03
N150N5	45.2	0.7	20.7	1.6	45.0	0.8	0.3	0.03
N200N1	64.3	0.6	33.3	2.2	64.5	0.8	0.3	0.03
N200N2	60.1	0.7	36.2	2.6	60.5	0.5	0.4	0.04
N200N3	61.3	0.6	35.8	2.4	61.2	0.7	0.4	0.03
N200N4	66.1	1.1	33.1	3.3	66.0	1.4	0.3	0.06
N200N5	62.8	0.4	35.0	1.5	62.5	0.9	0.3	0.04
N250N1	76.3	0.8	52.8	4.6	76.9	0.7	0.4	0.03
N250N2	76.2	1.0	53.5	3.0	76.2	1.1	0.5	0.04
N250N3	77.6	0.7	51.8	3.4	77.5	0.7	0.5	0.02
N250N4	73.4	0.7	54.8	3.9	73.9	0.5	0.4	0.05
N250N5	77.3	0.6	52.3	4.2	77.3	0.6	0.5	0.06

Table 9.2: Numerical results for the independent set problem.

and $O(|V|^{1.4})$, respectively. For the independent set problem, they are estimated to be $O(|V|^{1.7})$ and $O(|V|^{1.1})$, respectively. For the graph colouring problem, they are estimated to be $O(|V|^{1.6})$ and $O(\frac{|V|^{1.1}}{\Delta+1})$, respectively. As the time needed to carry out a trial is constant for both algorithms the time complexities do not depend on the exact implementation of the algorithms.

We mention that the Boltzmann machine uses a conceptually simple cooling schedule whereas simulated annealing uses a more elaborate one. By tuning the cooling schedule of the Boltzmann machine, one might even improve the computation times of the Boltzmann machine.

Problem	Simulated Annealing				Boltzmann Machine			
instance	$\overline{f}$	σ_f	$\overline{t}$	σ_t	$\overline{C}$	σ_C	$\overline{t}$	σ_t
N50N1	6.2	0.6	83.3	3.8	6.1	0.3	0.6	0.03
N50N2	5.3	0.5	82.7	2.9	5.9	0.7	0.5	0.07
N50N3	6.0	0.0	82.5	2.5	5.8	0.4	0.6	0.04
N50N4	6.4	0.5	84.1	3.5	6.1	0.3	0.5	0.07
N50N5	6.2	0.4	81.6	3.5	6.1	0.3	0.5	0.07
N100N1	6.0	0.0	243.3	8.5	6.2	0.4	1.2	0.06
N100N2	5.6	0.5	234.7	4.6	6.1	0.5	1.2	0.05
N100N3	5.7	0.5	241.6	4.9	5.6	0.5	1.2	0.06
N100N4	5.8	0.4	241.0	5.2	6.1	0.3	1.2	0.06
N100N5	5.9	0.3	239.0	4.4	5.4	0.5	1.2	0.04
N150N1	5.9	0.3	443.6	8.0	5.7	0.5	1.9	0.07
N150N2	6.0	0.0	446.7	8.7	5.8	0.4	1.9	0.09
N150N3	6.0	0.0	449.7	14.5	6.3	0.5	1.9	0.08
N150N4	5.9	0.3	452.6	11.6	5.7	0.5	1.8	0.08
N150N5	5.9	0.3	450.4	11.9	5.9	0.7	1.8	0.12
N200N1	5.9	0.3	686.7	9.9	5.8	0.4	2.5	0.07
N200N2	5.9	0.3	696.2	12.5	6.1	0.3	2.6	0.06
N200N3	5.9	0.3	684.0	11.6	5.9	0.7	2.6	0.12
N200N4	5.9	0.3	685.1	14.3	5.6	0.5	2.5	0.05
N200N5	5.8	0.4	701.9	14.5	5.6	0.5	2.5	0.05
N250N1	6.0	0.4	997.4	25.5	5.8	0.4	3.3	0.10
N250N2	6.0	0.0	983.2	15.5	6.2	0.4	3.3	0.12
N250N3	5.9	0.3	990.9	13.0	6.0	0.0	3.2	0.10
N250N4	6.0	0.0	990.0	22.3	6.1	0.3	3.2	0.08
N250N5	6.0	0.0	991.1	17.5	5.9	0.7	3.2	0.11

Table 9.3: Numerical results for the graph colouring problem.

9.7.2 The Travelling Salesman Problem

Computer simulations of the Boltzmann machine for the travelling salesman problem are carried out for two instances with 10 and 30 cities, respectively. Due to the large computation times required by the Boltzmann machine we have restricted ourselves to these small problem instances. The Boltzmann machine simulations were carried out in a similar way as was done for the graph problems discussed in Section 9.7.1. The simulations, carried out on a VAX 11/785, required computation times ranging from a few minutes, for the 10 cities problem instance, up to a few hours, for the 30 cities problem instance.

	10 cities	30 cities
$\bar{\ell}$	2.815	5.459
σ	0.141	0.312
$\ell_\vee$	2.675	4.929
$\ell^\wedge$	3.277	6.044
M	100	25
I	1.2	1.4
ℓ_{min}	2.675	4.299

Table 9.4: Numerical results obtained for instances of the travelling salesman problem with 10 and 30 cities, respectively (for explanation of the symbols see text).

The quality of the final solutions obtained by the simulations is investigated by collecting statistics. Table 9.4 shows the results of a statistical analysis for the two problem instances taken from Hopfield & Tank [1985]. The samples are obtained by running the computer simulation programs M times for different initial configurations. The various quantities used in the table are defined as follows:

$\bar{\ell}$: average tour length,
σ : spreading in the tour length,
$\ell_\vee$: smallest observed tour length,
$\ell^\wedge$: largest observed tour length,
M : sample size,
I : average number of iterations, and
ℓ_{min} : smallest known value of the tour length.

The notion 'tour length' refers to the length of the final tour obtained by the simulations. As a result of the special choice of the connection strengths discussed above, the feasibility of C does not hold for the complete region of values of the consensus function, but it only holds in the near-optimal region. Consequently, consensus maximization may yield a final solution that corresponds to a non-tour. If a non-tour is obtained at the end of the consensus maximization, then the maximization algorithm is restarted. The average number of iterations required to obtain a tour is denoted by I. From the simulations, we conclude that for the travelling salesman problem the Boltzmann machine cannot obtain results that are comparable to the results obtained by simulated annealing. For the problem instances used here the simulated annealing algorithm finds optimal solutions in a few seconds. An explanation of this phenomenon is given in Section 9.8. Specifying the travelling salesman problem as a linear assignment problem leads to comparable results when implemented on a Boltzmann machine [Aarts & Korst, 1988a].

We end this section with some observations. In an important work Hopfield & Tank [1985] introduced neural networks where the computing elements are linear analog neurons. They showed through computer simulations that near-optimal solutions for combinatorial optimization problems (e.g. the travelling salesman problem) could be obtained with these networks. Typical values obtained by Hopfield and Tank for the problem instances discussed above are 2.82 for the 10-cities problem and 5.65 for the 30-cities problem. The networks used by Hopfield and Tank are completely connected, i.e. each individual computing element is connected to all other elements. For a travelling salesman problem with n cities $O(n^4)$ connections are required. The typical nature of the analog computing elements hampers the implementation of these networks on general-purpose massively parallel computer systems such as for example the connection machine [Hillis, 1985]. However, one should take into account that their approach is not designed for implementation on such computer systems, but on special connectionist network architectures using dedicated hardware. Furthermore, we note that the typical analog nature of the computing elements, which Hopfield and Tank find to be essential for obtaining satisfactory results with their networks, is circumvented in our approach by the stochastic state transition mechanism.

9.8 Discussion

The architectural complexity of the Boltzmann machine for a number of graph problems is indicated in Table 9.5. For a travelling salesman problem with n cities the required numbers of units and connections are given by n^2 and $2n^3 - n^2$, respectively. For the graph problems discussed above, the Boltzmann machine performs very well, i.e. it converges quickly to near-optimal results. Final solutions can be obtained by the Boltzmann machine which are comparable, in quality, to the solutions obtained by simulated annealing. For all three problems the Boltzmann machine uses considerably less computation time, provided the model is emulated on a parallel computer.

From the simulations discussed in Section 9.7.2, we conclude that it is

Problem	Number of units	Number of connections
Max cut	$\mathcal{O}(n)$	$\mathcal{O}(n+m)$
Independent set	$\mathcal{O}(n)$	$\mathcal{O}(n+m)$
Vertex cover	$\mathcal{O}(n)$	$\mathcal{O}(n+m)$
Clique	$\mathcal{O}(n)$	$\mathcal{O}(n^2-m)$
Graph colouring	$\mathcal{O}(n(\Delta+1))$	$\mathcal{O}(n\Delta^2+m\Delta)$
Clique partitioning	$\mathcal{O}(n(\Delta+1))$	$\mathcal{O}(n\Delta^2+\Delta(n^2-m))$

Table 9.5: The architectural complexity of the Boltzmann machine for various graph problems, where n denotes the number of vertices, m the number of edges, and Δ the maximal degree of the given graph.

much harder to obtain near-optimal results for the travelling salesman problem than for the graph problems discussed in this paper. Two reasons can be given to explain this feature.

Firstly, choosing appropriate connections strengths for the travelling salesman problem is difficult. If the strengths are chosen such that the consensus function is feasible, then the convergence of the Boltzmann machine is slow due to the fact that the difference in consensus between good and bad tours is relatively small as compared to the differences in consensus between tours and non-tours. Choosing the connection strengths such that these differences are large, results in a situation where final results are often infeasible.

Secondly, transfering a given tour into another one, using the given Boltzmann machine formulation, requires in many cases a number of steps in which configurations are visited corresponding to infeasible solutions, which often have low values of the consensus. This increases the probability of getting trapped in configurations corresponding to locally optimal tours whose length deviates substantially from that of an optimal tour.

Evidently, both reasons given above, strongly depend on the construction that is used for the travelling salesman problem in the Boltzmann machine. It is however hard to think of other, more efficient constructions for this problem.

The importance of the Boltzmann machine as a massively parallel implementation of simulated annealing becomes more significant when Boltzmann machines are implemented on special-purpose hardware or general neurocomputers [Treleaven, 1988]. In this way the annealing process can be performed extremely fast by using analog devices for instance, which add up the incoming charge and make the stochastic decisions by means of noise. The design of hardwired networks has been the subject of study for some time. Recently, Alspector & Allen [1987] presented a design of a VLSI chip with $5 \cdot 10^5$ gates, implementing a Boltzmann machine consisting of approximately 2000 units (this design is also suitable for learning tasks). They estimate that their chip will run about a million times faster than simulations on a VAX. Optical implementations of the Boltzmann machine such as proposed by Farhat [1987] and Ticknor & Barrett [1987] are claimed to increase this factor even further.

Interest in solving combinatorial optimization problems with neural networks is growing rapidly. Among the large number of papers that have been presently published in the literature on this subject, we mention the following applications: graph colouring [Dahl, 1987], scheduling [Gulati, Iyengar, Toomarian, Protopopescu & Bahren, 1987], travelling salesman problem [Bagherzadeh, Kerola, Leddy & Brice, 1987; Gutzmann, 1987; Kamgar-Parsi & Kamgar-Parsi, 1987], min-cut [Bruck & Goodman, 1987], global optimization [Levy & Adams, 1987], non-attacking queens problem and detection of graph isomorphism [Tagliarini & Page, 1987]. Not all of these papers are based on the Boltzmann machine, but the various approaches presented by the authors can be straightforwardly formulated within this model.

CHAPTER 10

Classification and Boltzmann Machines

The taxonomy of applications presented in Section 8.4 distinguishes between two main classes of problems in neural computing, viz. search problems and learning problems. In Chapter 9 we demonstrated the use of Boltzmann machines for solving (approximately) a number of combinatorial optimization problems, which can be viewed as a subclass of search problems. In this chapter we focus our attention on another subclass, viz. classification problems.

Classification problems originate from the field of *pattern recognition* and constitute a class of problems that can often be easily solved by human beings, but are very hard to solve by computers [Devijver & Kittler, 1987; Young & Calvert, 1974]. Conventional computers, based on the Von Neumann architecture, perform substantially less well than 'biological pattern recognizers', especially in areas such as speech and image recognition. Recently, the interest in connectionist network models has considerably increased since there is a general belief that they are more suitable for solving these classification problems [Ballard, Hinton & Sejnowski, 1983; Feldman & Ballard, 1982; Hinton & Anderson, 1981; Sejnowski & Hinton, 1985].

In this chapter we discuss the use of Boltzmann machines for solving classification problems as an introduction to Chapter 11, where we extensively discuss the subject of learning with a Boltzmann machine. This is a natural approach, since considering a classification problem as a search problem is a first step towards considering it as a learning problem.

Before we give a formal definition of classification problems, we briefly consider the analogy between the functioning of a human brain and a Boltzmann machine; see also Chapter 7.

From an abstract point of view, a brain can be viewed as a network of neurons, storing information in the strengths of the synaptic connections [Hebb,

1949]. In perception, a subset of the neurons of the brain receives external stimuli, for instance from the visual system. The neurons then adjust their states according to the incoming stimuli and the previously stored information. In this way the network responds to the given external stimuli.

A Boltzmann machine can be modelled to operate in a similar way. To this end, the external stimuli are considered to clamp the states of a subset of the units. Then, given a connection pattern and a set of connection strengths, the remaining free units adjust their states in order to maximize the consensus, subject to the states of the clamped units. By doing so, the Boltzmann machine constructs the most probable configuration relative to the states of the clamped units, based on the information that is stored in the connection strengths.

Complex classification problems that are of practical interest in the field of pattern recognition usually involve a number of preprocessing steps, for instance concerning noise detection and data reduction; see Devijver & Kittler [1987] and Young & Calvert [1974]. We do not deal with these preprocessing steps here, but we focus our attention on the central problem, i.e. the classification itself, and we show how Boltzmann machines can be used to solve this problem.

10.1 Classification Problems

A classification problem can be formalized as a pair (O, S) where O denotes a finite set of objects $O = \{o_1, \ldots, o_n\}$, and S denotes a collection of disjoint subsets $S_1, \ldots, S_m$ of O. The problem is to classify automatically a given object $o_i \in O$ as a member of one of the subsets $S_j \subset S$.

For example, consider the problem of automatically recognizing images of a single handwritten character as one of the 26 letters in the Roman alphabet. Then the set of objects O can be defined as the set of all possible images consisting of $n \times n$ binary pixels, and the subsets $S_j, j = 1, \ldots, 26$ can be defined as the sets of all possible images that correspond to the letter j.

Clearly, many of the $n \times n$ images cannot be interpreted as meaningful, i.e. $O \neq \bigcup_{j=1}^{26} S_j$, since they do not correspond to a letter in the alphabet. There are generally two possible approaches to deal with such a situation.

(1) Extend the set of subsets with a subset S_{m+1} containing all 'meaningless' images, such that the classification recognizes a meaningless or ambiguous image as such.

(2) Force the classification to interpret all possible images as meaningful, and provide each outcome with a measure indicating the reliability of

the classification, such that whenever an image is ambiguous or less meaningful this measure indicates a low reliability.

Classification strategies based on approach (1) require a complete and explicit specification of all subsets. In practice, however, the set of objects is usually very large, and providing an explicit description of each subset, for instance by an enumeration of all its elements, is impracticable. Consequently, one is forced to use an implicit description of each subset S_j, which is usually done by specifying a number of 'typical' elements for each subset. If such an implicit description is used, the classification strategy should be based on approach (2). Furthermore, we mention that approach (2) presumes the use of associative capabilities, which is generally considered as a characteristic feature of neural networks such as the Boltzmann machine.

A Boltzmann machine can be used as a classification model in the following way. The states of a subset of the units in a Boltzmann machine, the *input units*, are clamped to some input pattern. This input pattern is a coded representation of an object o_i which is to be classified. The remaining *free units* then adjust their states to maximize the consensus, subject to the states of the input units and a given set of connection strengths. After maximization of the consensus, the states of a subset of the free units, i.e. the *output units*, represent the subset S_j to which o_i is thought to belong. The value of the consensus obtained for a given output can be used as a measure for the reliability of the classification. In this way, the Boltzmann machine is used as an input-output model which gives for each input a corresponding output and which gives an indication of the reliability of the input-output combination. Thus, we aim at using approach (2).

Evidently, using a Boltzmann machine for classification poses two problems that must be solved.

(i) Find an appropriate structure, i.e. a set of units and connections between the units.

(ii) Find an appropriate set of connection strengths.

We come back to these two problems later.

10.2 Extension of the Structural Description

To use a Boltzmann machine as an input-output model we need to extend the structural description given in Section 8.2. For this, the set of units is divided into three disjoint subsets, U_i, U_h and U_o, denoting the sets of *input, hidden* and *output units*, respectively. The set $\mathcal{U}_{io} = \mathcal{U}_i \cup \mathcal{U}_o$ denotes the set of *environmental units*.

Definition 10.1 An *input configuration* l of a Boltzmann machine is a global state of the input units and is uniquely defined by a sequence of length $|\mathcal{U}_i|$, whose u^{th} component $l(u)$ denotes the *state* of an input unit u in the input configuration l. The *input configuration space* $\mathcal{I}$ is given by the set of all possible input configurations. Clearly, the cardinality of $\mathcal{I}$ equals $2^{|\mathcal{U}_i|}$. ∎

Definition 10.2 An *output configuration* l of a Boltzmann machine is a global state of the output units and is uniquely defined by a sequence of length $|\mathcal{U}_o|$, whose u^{th} component $l(u)$ denotes the *state* of an output unit u in the output configuration l. The *output configuration space* $\mathcal{O}$ is given by the set of all possible output configurations. ∎

Input and output configurations are often simply denoted by inputs and outputs, respectively.

Definition 10.3 An *environmental configuration* l of a Boltzmann machine is a global state of the environmental units and is uniquely defined by a sequence of length $|\mathcal{U}_{io}|$, whose u^{th} component $l(u)$ denotes the *state* of an environmental unit u in the environmental configuration l. The *environmental configuration space* $\mathcal{V}$ is given by the set of all possible environmental configurations. Clearly, we have $\mathcal{V} = \mathcal{I} \times \mathcal{O}$. ∎

Definition 10.4 An *internal configuration* l of a Boltzmann machine is a global state of the hidden units and is uniquely defined by a sequence of length $|\mathcal{U}_h|$, whose u^{th} component $l(u)$ denotes the *state* of a hidden unit u in the internal configuration l. The *internal configuration space* $\mathcal{H}$ is given by the set of all possible internal configurations. ∎

With each input configuration l a subspace $\mathcal{R}_{I(l)}$ of configurations $k \in \mathcal{R}$ can be associated given by

$$\mathcal{R}_{I(l)} = \{k \in \mathcal{R} \mid \forall v \in \mathcal{U}_i : k(v) = l(v)\}, \tag{10.1}$$

i.e. $\mathcal{R}_{I(l)}$ consists of all configurations $k \in \mathcal{R}$ for which the states of the input units are given by l. The subspaces $\mathcal{R}_{O(l)}$ and $\mathcal{R}_{V(l)}$ for the output configurations and the environmental configurations, respectively, are defined similarly.

Furthermore, consider two arbitrary (sub)spaces $\mathcal{A}$ and $\mathcal{B}$ of $\mathcal{R}$ with $|\mathcal{A}| > |\mathcal{B}|$, and let $k \in \mathcal{A}$. Then $\mathcal{L}_{\mathcal{B}}(k)$ denotes the *projection* of k onto the subspace $\mathcal{B}$. In other words, $\mathcal{L}_{\mathcal{B}}(k)$ denotes the part of k that is contained in $\mathcal{B}$. For instance let $k \in \mathcal{V}$, then $\mathcal{L}_{I}(k)$ denotes the input part of the environmental configuration k.

Now, we introduce a *classification set* $\mathcal{V}' \subseteq \mathcal{V}$, which specifies a number of objects o_i and the corresponding classification sets S_j in terms of input and output configurations, respectively. The corresponding input and output parts are denoted by $\mathcal{I}'$ and $\mathcal{O}'$, respectively.

Classification in a Boltzmann machine now amounts to finding a correct output for a given input. As mentioned before this is done by constrained maximization of the consensus function.

Before demonstrating the use of a Boltzmann machine for classification, we make some remarks about the relationship between classification problems and optimization problems. Clearly, both classes of problems can be solved with a Boltzmann machine by maximization of the consensus as discussed in Chapter 8. In solving optimization problems, all units of a Boltzmann machine are allowed to adjust their state in order to maximize the consensus. In solving classification problems, the states of the input units are clamped and only a subset of the units are allowed to adjust their states. In this case, the Boltzmann machine can only assume a subset of configurations in $\mathcal{R}$, determined by the given input. Each input determines a different subset of $\mathcal{R}$. Thus, a classification problem can be considered as a collection of optimization problems, each defined on a different subset of $\mathcal{R}$. The different optimization problems, constituting the classification problem, give rise to different and potentially conflicting requirements on the connection strengths. Therefore, finding appropriate connection strengths is generally much harder for classification problems than for optimization problems.

10.3 Examples

The use of Boltzmann machines for classification is probably best explained by means of examples. In this section we discuss two examples which exhibit markedly different aspects, viz. the *seven-segment display problem* and the *exclusive-or problem*. The difference between both examples lies in the use of hidden units, i.e. for the seven-segment display, a set of non-conflicting connection strengths can be found and therefore the problem can be solved without using hidden units. The exclusive-or, however, gives rise to conflicting requirements on the set of connection strengths, which can only be dissolved by using hidden units.

10.3.1   Classification without Hidden Units

To discuss the problem of classification in the absence of hidden units, we use the following example.

Example 10.1 (Seven-segment display) For the display of decimal digits, for instance in a pocket calculator, one often uses a seven-segment display; see Figures 10.1a and 10.1b. Each of the seven segments in the display can be independently turned 'on' or 'off' in order to display one of the digits from 0 to 9. Assume that we want to classify the states of a seven-segment display as one of the 10 digits, according to the encoding given in Table 10.1.   As

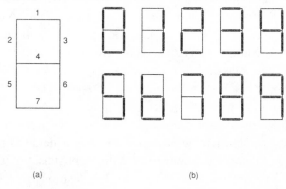

(a)   (b)

Figure 10.1: Schematic representation of the seven-segment display of the digits 0 - 9.

already mentioned, the problem is now to choose a Boltzmann machine with appropriate structure and connection strengths.

   With respect to the structure, the following choice is straightforward. An input unit is defined for each of the seven segments and an output unit for each of the ten digits. No hidden units are used. The classification set $\mathcal{V}'$ contains the ten environmental configurations given in Table 10.1, with input part $\mathcal{I}'$ and output part $\mathcal{O}'$.

   Since the states of each input and output unit are correlated, it is obvious to choose a connection pattern that connects each input unit with each output unit. Furthermore, as we are opting for approach (2), each possible input must be interpreted as meaningful. This implies that for each final configuration obtained by the Boltzmann machine, exactly one output unit should have state 1, while the other output units should have state 0. The latter can be achieved by introducing inhibitory connections between the output units.

   Thus, the structure of our Boltzmann machine is chosen to consist of 7 input units and 10 output units, while the set of connections is given by the

digit	input ($\mathcal{I}'$)	output ($\mathcal{O}'$)
0	1 1 1 0 1 1 1	1 0 0 0 0 0 0 0 0 0
1	0 0 1 0 0 1 0	0 1 0 0 0 0 0 0 0 0
2	1 0 1 1 1 0 1	0 0 1 0 0 0 0 0 0 0
3	1 0 1 1 0 1 1	0 0 0 1 0 0 0 0 0 0
4	0 1 1 1 0 1 0	0 0 0 0 1 0 0 0 0 0
5	1 1 0 1 0 1 1	0 0 0 0 0 1 0 0 0 0
6	0 1 0 1 1 1 1	0 0 0 0 0 0 1 0 0 0
7	1 0 1 0 0 1 0	0 0 0 0 0 0 0 1 0 0
8	1 1 1 1 1 1 1	0 0 0 0 0 0 0 0 1 0
9	1 1 1 1 0 1 0	0 0 0 0 0 0 0 0 0 1

Table 10.1: The environmental configurations that the Boltzmann machine must be able to recognize. Each ith column of the inputs corresponds to the ith segment and each ith column of the outputs corresponds to the ith digit.

union of the following two sets:

$$\mathcal{C}_{io} = \{\{u, v\} \mid u \in \mathcal{U}_i \wedge v \in \mathcal{U}_o\} \tag{10.2}$$

$$\mathcal{C}_{oo} = \{\{u, v\} \mid u \in \mathcal{U}_o \wedge v \in \mathcal{U}_o\}. \tag{10.3}$$

The corresponding connection pattern is illustrated in Figure 10.2. It goes without saying that we do not use connections between input units, since the states of the input units are always clamped. Furthermore, we do not use bias connections in this example.

Next, we must choose an appropriate set of connection strengths. For this, we adopt the following approach. A connection $\{u, v\} \in \mathcal{C}_{io}$ is either inhibitory or excitatory depending on the correlation between the states of input unit u and output units v for all configurations of the classification set $\mathcal{V}'$. More specifically, a connection $\{u, v\} \in \mathcal{C}_{io}$ is chosen excitatory if

$$\exists k \in \mathcal{V}' : k(u) \cdot k(v) = 1, \tag{10.4}$$

otherwise it is chosen inhibitory. As already mentioned, the strengths of the connections in $\mathcal{C}_{oo}$ should be chosen 'sufficiently' negative, such that at most one output unit is 'on' after consensus maximization. We can now formulate the following theorem.

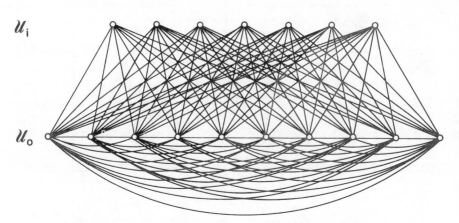

Figure 10.2: Connection pattern of the Boltzmann machine for the seven-segment display.

Theorem 10.1 *Let the connection pattern of the Boltzmann machine for the seven-segment display be given by (10.2) and (10.3), and let for all $v \in \mathcal{U}_o$ the sets $\mathcal{N}_v^+$, and $\mathcal{N}_v^-$ be defined as*

$$\mathcal{N}_v^+ = \{u \in \mathcal{U}_i \mid \{u, v\} \text{ is excitatory}\} \tag{10.5}$$

$$\mathcal{N}_v^- = \{u \in \mathcal{U}_i \mid \{u, v\} \text{ is inhibitory}\}. \tag{10.6}$$

Furthermore, let the connection strengths be given by

$$\forall u \in \mathcal{U}_i, \forall v \in \mathcal{U}_o : \quad s_{\{u,v\}} = \begin{cases} \frac{\gamma}{|\mathcal{N}_v^+|} & \text{if } u \in \mathcal{N}_v^+ \\[2mm] \frac{-\gamma}{|\mathcal{N}_v^-|} & \text{if } u \in \mathcal{N}_v^-, \end{cases} \tag{10.7}$$

and

$$\forall \{u, v\} \in \mathcal{C}_{oo} : \quad -\gamma < s_{\{u,v\}} < -\gamma + \epsilon, \tag{10.8}$$

where $\epsilon = \min_{u \in \mathcal{U}_o} \{\frac{\gamma}{|\mathcal{N}_u^+|}\}$ and γ some positive number, then

   (i) the consensus function C has exactly ten global maxima corresponding to the ten configurations of the classification set $\mathcal{V}'$; the value of a global maximum equals γ, and

   (ii) each local maximum of C contains an output part that belongs to the output part of the classification set. More specifically, let $\widehat{\mathcal{R}}$ denote the set of locally maximal configurations in $\mathcal{R}$, then

$$\forall k \in \widehat{\mathcal{R}} : \mathcal{L}_o(k) \in \mathcal{O}'. \tag{10.9}$$

Proof Part (i) can be straightforwardly proved by showing that the ten configurations of the classification set $\mathcal{V}'$ are the only configurations of the Boltzmann machine for which the sum of the strengths of the excitatory connections is maximal, whereas no inhibitory connections are activated. If no inhibitory connection is activated there is at most one output unit v having state 1, all other output units having state 0. Only in the trivial case, where all input units are clamped to state 0, it is possible that no output unit has state 1. In all other cases, exactly one output unit eventually has state 1. Then the maximal number of excitatory connections that can be activated equals $|\mathcal{N}_v^+|$, all having a strength equal to $\frac{\gamma}{|\mathcal{N}_v^+|}$. Hence, the maximal value of the consensus equals γ. Furthermore, the corresponding input configurations in this case equal the input parts of the configurations in the classification set $\mathcal{V}'$.

For the proof of part (ii), we recall that during consensus maximization the states of the input units are clamped to values of an input configuration corresponding to a given object that is to be classified, i.e. only the output units are allowed to change their states. Now, it is easy to show that for the choice of the connection strengths given in (10.7) and (10.8), the consensus can always be lowered if one of the inhibitory connections in $\mathcal{C}_{oo}$ is activated. Thus, each local maximum corresponds to a situation in which exactly one of the output units has state 1, while the other output units have state 0. Hence, the output part $\mathcal{L}_o(\hat{k})$ of each locally maximal configuration $\hat{k}$ equals one of the output configurations in $\mathcal{O}'$. ∎

From Theorem 10.1 it is evident that all objects in the classification set $\mathcal{V}'$ are classified correctly, part (i), and that any possible object at least is classified as meaningful, part (ii), which agrees with the requirements imposed by classification approach (2), discussed in Section 10.1.

To illustrate some features of the classification problem described above, we discuss some numerical results. Table 10.2 gives the strengths of the connections between the seven input units $u_1, \ldots, u_7$ and the ten output units $u_8, \ldots, u_{17}$ of the seven-segment display calculated according to (10.7) using $\gamma = 420$. Table 10.3 shows the value of the consensus of all possible combinations between the inputs and the outputs of the classification set $\mathcal{V}'$ of the seven-segment display. Indeed, we observe from the table that the value of the consensus is maximal if the input is classified correctly. To test the Boltzmann machine of the seven-segment display for its classification capabilities, a simulation program is implemented on a VAX 11/785. We use the parallel cooling schedule presented in Section 8.3.3, with the following parameters: $\alpha = 0.9, L = 10$ and $K = 10$. Furthermore, the strengths of the inhibitory

	u_8	u_9	u_{10}	u_{11}	u_{12}	u_{13}	u_{14}	u_{15}	u_{16}	u_{17}
u_1	70	-84	84	84	-140	84	-210	140	60	84
u_2	70	-84	-210	-210	105	84	84	-105	60	84
u_3	70	210	84	84	105	-210	-210	140	60	84
u_4	-420	-84	84	84	105	84	84	-105	60	84
u_5	70	-84	84	-210	-140	-210	84	-105	60	-210
u_6	70	210	-210	84	105	84	84	140	60	84
u_7	70	-84	84	84	-140	84	84	-105	60	-210

Table 10.2: Strengths of the connections between the input units and the output units of the Boltzmann machine for the seven-segment display.

connections $\{u, v\} \in C_{oo}$ are given the value -400.

Each of the ten inputs of the classification set $\mathcal{V}'$ are clamped 100 times to the states of the input units. The results are taken as the states of the output units obtained after consensus maximization. It is observed that the Boltzmann machine classifies all inputs in $\mathcal{V}'$ correctly all the time. This can be understood from the fact that even an iterative improvement algorithm finds the correct outputs for the inputs of the classification set, which can be easily verified from the values of the connection strengths.

To test the potential associative capabilities of the Boltzmann machine, the input units are clamped to inputs not belonging to the input part $\mathcal{I}'$ of classification set $\mathcal{V}'$. To illustrate the classification of these 'unknown' inputs we have chosen 10 inputs different from the ones given in Table 10.1. Again, each of these inputs is clamped to the input units 100 times. The results of this test are given in Table 10.4, where the leftmost column represents the inputs given to the Boltzmann machine. Table 10.5 again shows the value of the consensus of all possible combinations of the 'unknown' inputs and the outputs in $\mathcal{O}'$.

Table 10.4 clearly shows that the Boltzmann machine exhibits associative capabilities. For example, if the input corresponding to ⸮ is clamped to the input units, then the Boltzmann machine tends to interpret the input as a 2. More generally, if the input shows a great resemblance to some input in the classification set, then the probability of finding the corresponding output in the classification set is large.

Furthermore, from Tables 10.4 and 10.5 we observe that the frequency of

	0	1	2	3	4	5	6	7	8	9
0	420	84	-84	-84	-105	-84	-84	105	-60	-84
1	140	420	-126	186	210	-126	-126	280	120	-168
2	-140	-126	420	126	-210	-168	-168	-35	-120	-168
3	-140	168	126	420	35	126	-168	210	-120	126
4	-210	252	-252	42	420	42	42	70	-180	336
5	-140	-126	-168	126	35	420	126	-35	-120	126
6	-140	-126	-168	-168	35	126	420	-280	-120	-168
7	210	336	-42	252	70	-42	-336	420	180	252
8	0	0	0	0	0	0	0	0	420	0
9	-140	168	-168	126	280	126	-168	210	-120	420

Table 10.3: Consensus of all possible combinations of the inputs and outputs in the classification set of the seven-segment display.

	0	1	2	3	4	5	6	7	8	9
0	91	-	-	-	-	-	-	-	9	-
1	1	65	4	5	6	-	-	18	-	1
2	-	-	93	-	-	-	-	-	7	-
3	-	-	-	41	-	51	-	-	3	5
4	-	-	-	-	75	9	9	-	3	4
5	5	-	-	-	-	81	-	-	3	11
6	9	-	-	-	-	-	91	-	-	-
7	1	2	-	6	-	5	-	78	7	1
8	-	-	-	-	-	-	-	-	100	-
9	12	3	-	-	-	-	-	64	5	15

Table 10.4: Score matrix for the classification test with 10 'unknown' inputs.

	0	1	2	3	4	5	6	7	8	9
ᒋ	350	-126	-168	-168	-210	126	126	-35	300	-168
ᛁ	70	210	84	84	105	-210	-210	140	60	84
ᒣ	-210	-42	336	42	-70	-252	-252	70	240	42
ᒫ	-210	-42	42	336	-70	336	42	70	240	42
ᒧ	-280	42	-336	-42	315	252	252	-70	180	252
ᒲ	280	-42	-252	42	-70	336	42	70	240	42
ᒪ	280	-42	-252	-252	-70	42	336	-175	240	-252
ᛁ	140	126	-126	168	-35	168	-126	280	120	168
ᒷ	-70	84	210	210	-105	-84	-84	105	360	-84
ᒼ	280	252	-252	42	175	42	-252	315	240	336

Table 10.5: Consensus of all possible combinations of the 'unknown' inputs and outputs of Table 10.4.

obtaining an output strongly depends on the value of the consensus of the corresponding configuration. If the digits are recognized correctly the consensus is maximal, viz. $C = \gamma$. The lower the consensus, the lower the probability that the output is a correct interpretation of the input. Evidently, the value of the consensus function can be used as a reliability measure for the output found by the Boltzmann machine.   ∎

Finally, we remark that the procedure presented above for the Boltzmann machine of the seven-segment display can be easily generalized, i.e. one can always construct a Boltzmann machine without hidden units that correctly classifies n inputs, each representing a different class, if the number of outputs is chosen equal to n. In this case, each class can be associated with one of the output units, such that the requirements on the connection strengths for the n different inputs are independent, and consequently not conflicting. The corresponding values of the connection strengths can then be calculated from (10.7) and (10.8).

10.3.2   Classification with Hidden Units

In the previous section, it is shown that no hidden units were necessary to solve the classification problem associated with the seven-segment display; see Example 10.1. As pointed out, this is due to the fact that only one input is associated with each output in O', i.e. only one object o_i is associated with each classification subset S_j. Generally, we want to consider classification problems where a subset S_j contains several objects o_i, which can lead to conflicting requirements on the connection strengths; see Example 10.2. In that case, the approach presented in Section 10.3.1 can no longer be used.

The use of an extra layer of units, the hidden units, whose states can be freely chosen, introduces additional degrees of freedom that can be used to eliminate conflicting requirements on the connection strengths. This is a well-known approach in neural computing. Already in 1969 Minsky and Papert proved for the perceptron model that, in addition to an input and an output layer, an extra layer of hidden units was needed in order to obtain good results for many classification problems, including the well-known exclusive-or problem.

To illustrate the use of hidden units we discuss as an example the exclusive-or problem; see also Sejnowski, Kienker & Hinton [1986].

Example 10.2 (Exclusive-or problem) Suppose we have a Boltzmann machine with two input units, u_1 and u_2, and one output unit, u_3. The classification set $\mathcal{V}'$ is given in Figure 10.3a. The codes in the classification set are equivalent to the Boolean exclusive-or function. The configurations in the input part $\mathcal{I}'$ correspond to the states of the input units u_1 and u_2. The configurations in the output part $\mathcal{O}'$ corresponds to the states of the output unit u_3. From the fact that the configurations a, b, c, and d in Figure 10.3 need to be

k	input ($\mathcal{I}'$)	output ($\mathcal{O}'$)
a	0 0	0
b	0 1	1
c	1 0	1
d	1 1	0

(a)

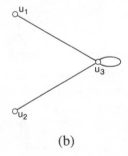

(b)

Figure 10.3: The inputs and the corresponding outputs for the exclusive-or problem (a), and the structure of the corresponding Boltzmann machine without hidden units (b).

local maxima of the consensus function we derive the following requirements on the connection strengths:

$$s_{\{u_3, u_3\}} \quad < \quad 0, \tag{10.10}$$

$$s_{\{u_2,u_3\}} + s_{\{u_3,u_3\}} \;>\; 0, \tag{10.11}$$

$$s_{\{u_1,u_3\}} + s_{\{u_3,u_3\}} \;>\; 0, \text{ and} \tag{10.12}$$

$$s_{\{u_1,u_3\}} + s_{\{u_2,u_3\}} + s_{\{u_3,u_3\}} \;<\; 0. \tag{10.13}$$

Each requirement corresponds to a row in Figure 10.3a. From (10.10) and (10.11) it follows that

$$s_{\{u_2,u_3\}} > 0. \tag{10.14}$$

Combining (10.12) and (10.14) we obtain

$$s_{\{u_1,u_3\}} + s_{\{u_2,u_3\}} + s_{\{u_3,u_3\}} > 0, \tag{10.15}$$

which clearly conflicts with the requirement of (10.13). Consequently, it is not possible to choose the connection strengths such that the Boltzmann machine is able to classify all four inputs correctly.

If the Boltzmann machine is extended with a hidden unit u_4 as shown in Figure 10.4b, and if the states of the hidden unit for the various environmental configurations of the classification set $\mathcal{V}'$ are chosen as indicated in Figure 10.4a, then the requirements on the connection strengths are no longer conflicting. It is easily verified that the following choice of the connection strengths leads to the desired classification behaviour:

$$s_{\{u_i,u_3\}} = 2, \qquad s_{\{u_i,u_4\}} = 6, \qquad s_{\{u_3,u_3\}} = -1,$$
$$s_{\{u_4,u_4\}} = -7, \qquad s_{\{u_3,u_4\}} = -4,$$

where $u_i \in \{u_1, u_2\}$.

The states of the hidden unit have been chosen such that a distinction is made between the environmental configurations that lead to the conflicting requirements on the connection strengths. More specifically, the states of the hidden unit introduce a distinction between the situation where both input units have state 1 and the situation where one of the input units has state 0. This distinction eliminates the conflicts in the requirements on the connection strengths for the exclusive-or. We note that this choice of the states of the hidden units is not unique; in fact there are 10 different choices for the internal configurations that lead to a correct classification; see Table 11.2.   ∎

In general terms, the role of the hidden units is to constitute an additional layer of units that can be used to encode a number of typical features of the relations between the environmental configurations. Basically, each classification problem can be solved by using a sufficiently large number of hidden units. This is formalized by the following theorem.

input (I')	hidden (H')	output (O')
0 0	0	0
0 1	0	1
1 0	0	1
1 1	1	0

(a)

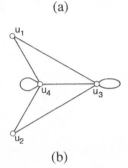

(b)

Figure 10.4: The inputs and outputs for the exclusive-or problem extended with one hidden unit, whose states are chosen such that the requirements on the connection strengths do not conflict (a); the structure of the corresponding Boltzmann machine with one hidden unit u_4 (b).

Theorem 10.2 *Let a classification problem be specified by a finite classification set $\mathcal{V}'$ of a Boltzmann machine with $|U_i|$ input units and $|U_o|$ output units. Furthermore, let each environmental configuration in $\mathcal{V}'$ have a unique input part. Then the connection strengths can always be chosen such that the Boltzmann machine generates for each input in $\mathcal{V}'$ the corresponding output, if the number of hidden units $|U_h|$ is chosen equal to $|\mathcal{V}'|$.*

Proof By choosing $|U_h|$ equal to $|\mathcal{V}'|$, it is possible to associate with each environmental configuration $i \in \mathcal{V}'$ a unique hidden unit u_i, such that unit u_i eventually has state 1 if the input units are fixed to the input corresponding to environmental configuration i. Thus, with each environmental configuration i, an internal configuration k is associated such that $k(u_i) = 1$ and $k(u_j) = 0$, for all $u_j \in U_h \backslash \{u_i\}$. Consequently, the requirements imposed on the connection strengths by the different environmental configurations are independent. The connection between a hidden unit u_i and some environmental unit u_j is chosen to be excitatory if $i(u_j) = 1$, and inhibitory otherwise.

Let the set of connections be given by the union of the following three sets:

$$C_{ih} = \{\{u, v\} \mid u \in \mathcal{U}_i \wedge v \in \mathcal{U}_h\}, \tag{10.16}$$

$$C_{hh} = \{\{u, v\} \mid u, v \in \mathcal{U}_h \wedge u \neq v\}, \quad \text{and} \tag{10.17}$$

$$C_{ho} = \{\{u, v\} \mid u \in \mathcal{U}_h \wedge v \in \mathcal{U}_o\}, \tag{10.18}$$

and let the connection strengths be given by the following relations:

$$\forall v \in \mathcal{U}_h, \forall u \in \mathcal{N}_v^+ \quad : \quad s_{\{u,v\}} = \frac{\Gamma}{|\mathcal{N}_v^+|}, \tag{10.19}$$

$$\forall v \in \mathcal{U}_h, \forall u \in \mathcal{N}_v^- \quad : \quad s_{\{u,v\}} = \frac{-\Gamma}{|\mathcal{N}_v^-|}, \tag{10.20}$$

$$\forall \{u, v\} \in C_{hh} \quad : \quad \gamma - \Gamma < s_{\{u,v\}} < \epsilon - \Gamma - \gamma, \tag{10.21}$$

$$\forall u \in \mathcal{U}_h, \forall v \in \mathcal{M}_i^+ \quad : \quad s_{\{u,v\}} = \frac{\gamma}{|\mathcal{M}_u^+|}, \quad \text{and} \tag{10.22}$$

$$\forall u \in \mathcal{U}_h, \forall v \in \mathcal{M}_i^- \quad : \quad s_{\{u,v\}} = \frac{-\gamma}{|\mathcal{M}_u^-|}, \tag{10.23}$$

where $\mathcal{N}_v^+$ and $\mathcal{N}_v^-$ denote the sets of input units that have an excitatory and inhibitory connection with hidden unit v, respectively. Similarly, $\mathcal{M}_u^+$ and $\mathcal{M}_u^-$ denote the sets of output units that have an excitatory and inhibitory connection with hidden unit u, respectively. Γ, γ, and ϵ denote positive numbers with $\epsilon = \min_{u \in \mathcal{U}_h}\{\frac{\Gamma}{|\mathcal{N}_u^+|}\}$ and $\gamma < \frac{\epsilon}{2}$. Then all input parts in $\mathcal{V}'$ can be correctly classified.

The correctness of this statement can now be proved along the following lines:

- Firstly, it can be proved that - under the condition that one of the inputs in $\mathcal{I}'$ is clamped to the input units - each local maximum of the consensus function corresponds to a configuration in which exactly one hidden unit has state 1, the other hidden units having state 0.

- Next, it can be proved that the correct hidden unit eventually has state 1 if an input in $\mathcal{I}'$ is clamped to the input units. This proof is independent of the states of the output units.

- Finally, it can be proved that the correct output configuration is found provided the correct hidden unit has state 1.

A detailed proof of the theorem is left to the reader.   ∎

For large classification problems, the approach described by Theorem 10.2 is clearly not desirable, since the number of hidden units grows exponentially with the number of inputs. Furthermore, it should be noted that the number of internal configurations associated with an environmental configuration is very small in the approach described by Theorem 10.2. From the $2^{|\mathcal{U}_h|}$ possible internal configurations only $|\mathcal{U}_h|$ configurations are actually used as local maxima. In many cases the internal configurations can be used more efficiently, such that fewer hidden units are needed. Given that input and output units are not directly connected, and given that each input in $\mathcal{I}'$ corresponds to a unique output in $\mathcal{O}'$, then we have to associate with each environmental configuration in $\mathcal{V}'$ a unique internal configuration, in order to be able to implement the required behaviour. In that case, a lower bound on the required number of hidden units is given by $|\mathcal{U}_h| \geq \lceil\, ^2\log|\mathcal{V}'|\rceil$, in which case each internal configuration is associated with an environmental configuration in $\mathcal{V}'$. A major drawback however is that it is not always possible the find a corresponding set of connection strengths. In other words the internal configurations that go with the minimal number of hidden units may give rise to conflicting requirements on the connection strengths. We deal with this subject in the next section.

The use of hidden units requires the specification of (i) internal configurations and (ii) a set of connection strengths. As already mentioned, no direct constructive approach is known for deriving the required specifications in the general case. However, in Chapter 11 it will be shown that it is possible to augment the Boltzmann machine with a learning algorithm such that both specification problems are solved simultaneously for a Boltzmann machine with a fixed structure and a given classification set.

10.4   Discussion

We end this chapter with a discussion of the potential benefits and some open problems of Boltzmann machines - or connectionist models in general - for solving problems in the field of pattern recognition. First, we discuss two properties of the Boltzmann machine which make the model especially interesting for applications in the field of pattern recognition.

10.4.1 Association

As we already mentioned in Section 10.1, many practical problems within the field of pattern recognition cannot be well defined. If the number of objects is large, then an explicit description of the different classes used in the classification is impracticable for most problems. For example, if the set of inputs is given by the $n \times n$ binary images of human faces, then an explicit description of the classes 'male' and 'female' is virtually impossible.

In the discussion of Example 10.1, we have seen that a Boltzmann machine has associative capabilities. By this we mean that the model can classify an 'unknown' object as the 'known' object that resembles it best. This is especially useful for problems that are difficult to describe explicitly. If we make sure that a Boltzmann machine correctly classifies a number of representative or typical examples, the model may use its associative capabilities to classify inputs that resemble the given examples. Instead of giving an explicit description of the classification problem, it suffices to give a number of 'typical' examples, and the corresponding required outputs. In the same discussion we have also seen that the consensus can be used as a measure of reliability.

10.4.2 Fault Tolerance

Large classification problems can usually only be solved at the expense of large numbers of hidden units [Prager, Harrison & Fallside, 1986]. If a distributed internal representation is used, then one environmental configuration involves many activated connections and one connection is activated for many environmental configurations. This makes a hardware implementation of the Boltzmann machine sufficiently robust to withstand minor physical damage such as short-circuits in connections or malfunctioning units (fault tolerance).

This property of fault tolerance is generally considered as one of the most attractive features of neurocomputers compared to conventional computers. The differences between neurocomputers and conventional computers are, in this respect, analogous to the differences between holograms and photographs. For instance, cutting away a part of a photograph results in the loss of local information, that might be vital, while the remainder of the photograph is not changed. Cutting away a small part of the hologram does not result in the loss of some specific local information, but it makes the total picture more vague, as if it were a little out of focus.

Similarly, minor physical damage in a Boltzmann machine doesn't necessarily mean that a specific input-output combination is lost but it might result in a slight deterioration of its overall performance.

The subject of fault tolerance of Boltzmann machines is also addressed by Derthick [1984]. We return to this subject in the next section and in Section 11.7.

10.4.3 Open problems

The properties of the Boltzmann machine discussed in the previous sections illustrate the potential benefits of connectionist models, such as the Boltzmann machine, within the field of pattern recognition.

There are, however, a number of problems that have hardly been touched upon in the literature, though solving these problems would be of great importance to future developments in neural computing. In this section we formulate three such problems, none of which has been solved in a general sense. The problems are formulated with the Boltzmann machine in mind, but related problems also hold for other connectionist models.

In order to use the Boltzmann machine as a classifier it is important to know

- how many (hidden) units are needed,

- how the units should be interconnected, and

- how the connection strengths should be chosen.

These problems are treated in greater detail below.

Number of Hidden Units
The minimal number of hidden units required to solve a specific classification problem strongly depends on the intrinsic difficulty of the problem at hand. Theorem 10.2 gives a general approach to choosing the number of hidden units and the appropriate connection strengths. However, for this general approach, the number of hidden units grows exponentially with the number of input units, which is undesirable if we want to solve large classification problems.

Minsky & Papert [1969] showed that a number of notorious problems can only be solved by connectionist models if the number of hidden units grows exponentially with the number of input units.

However, for many practical applications one can do with fewer hidden units. If a classification problem has a highly regular structure, then the number of hidden units may be small; cf. the encoder problems discussed by Hinton, Sejnowski & Ackley [1984].

As observed earlier, the general approach given by Theorem 10.2 uses only a very small fraction of the total number of available internal configurations for representing an environmental configuration. Instead of using one hidden unit to represent one environmental configuration, we can use a distributed representation where in principle each internal configuration can be used to represent an environmental configuration. Consequently, the number of hidden units required for a specific classification problem is much smaller.

The actual minimal number of hidden units is given by the set of units for which it is possible to assign an internal configuration to each environmental configuration in the classification set, such that the resulting requirements on the connection strengths are not conflicting and the total number of hidden units is minimal. Unfortunately, no general rule is known from which this set can be calculated.

In many cases, however, the number of hidden units is taken to be somewhat larger than this minimum, for two reasons. Using more hidden units than strictly necessary simplifies the problem of finding an appropriate internal configuration space and corresponding set of connection strengths, and increases the fault tolerance, since the Hamming distance between the internal configurations may be increased.

Choice of Connection Pattern

Given a fixed number of input units, hidden units and output units, the problem is to find an appropriate set of connections between the units. Clearly, the most general solution is obtained by connecting all units with each other, since every required connection pattern then can be emulated on this complete interconnection pattern by choosing a subset of the connection strengths to be equal to 0. However, complete interconnection patterns hinder special-purpose hardware implementation in VLSI, or emulation on general-purpose neural computers. Furthermore, they may reduce the efficiency of the Boltzmann machine.

Basically, one may assume that units whose states are uncorrelated need not be connected. This leads to the following commonly used connection pattern for the Boltzmann machine when applied to classification problems [Aarts & Korst, 1986; 1988b; Ackley, Hinton & Sejnowski, 1985; Hinton, Sejnowski & Ackley, 1984]:

- all units have a bias connection,
- all input units are interconnected,
- all output units are interconnected,
- all input units are connected to all hidden units, and
- all output units are connected to all hidden units.

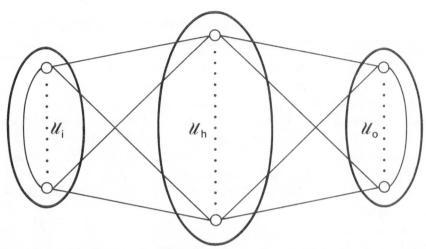

Figure 10.5: Schematic of the Boltzmann machine connection pattern commonly used for solving classification problems.

This connection pattern is illustrated by Figure 10.5. For large practical problems one could decide to introduce more structure into the internal part of the Boltzmann machine by choosing a multilayered structure for the hidden units. In this way, a hierarchical classification could be used where hidden units in a lower layer, i.e. a layer closer to the input units, are used to detect low-level features that are used by hidden units of a higher layer to detect more global features. Such a hierarchical structure, which greatly reduces the complexity of large problems, also seems to be present in the human visual system [Ballard, Hinton & Sejnowski, 1983; Feldman, 1981].

Choice of Connection Strengths
For a classification problem specified by a given classification set $\mathcal{V}'$, it is always possible to construct an appropriate structure and corresponding connection strengths such that the Boltzmann machine classifies all inputs in $\mathcal{V}'$ correctly, provided the set of hidden units is large enough, i.e. $|\mathcal{U}_h| = |\mathcal{V}'|$; see Theorem 10.2. However, as already mentioned, no simple rules are known for constructing a Boltzmann machine if the number of hidden units is smaller than this bound. The problem is that the states of the hidden units are not directly determined by the given input-output combinations. More specifically, the problem is to

- find for each environmental configuration in $\mathcal{V}'$ an appropriate internal configuration, such that the combined configurations lead to a set $\mathcal{R}' \subset \mathcal{R}$, and to

- find a set of connection strengths for which $\mathcal{R}'$ constitutes the set of (global) maxima of the consensus function.

An interesting approach for choosing the connection strengths such that a given subset $\mathcal{R}'$ constitutes the set of global maxima of the consensus function is proposed by Ten Bosch [1988]. The approach is based on the topology of the configurations in $\mathcal{R}'$, when viewed as vertices of an n-dimensional hypercube $\{0, 1\}^n$, where n denotes the number of units. We restrict ourselves to illustrating the approach by means of an example. Evidently, this approach presumes the presence of an internal representation, i.e. a choice of the internal configurations. For a detailed description the reader is referred to Ten Bosch [1988].

Suppose the subset $\mathcal{R}'$ consists of the following four configurations: $\mathbf{x} = (0001)$, $\mathbf{y} = (0110)$, $\mathbf{z} = (1010)$, and $\mathbf{w} = (1100)$. Now, the connection strengths can be chosen as follows. The four points $\mathbf{x}$, $\mathbf{y}$, $\mathbf{z}$, and $\mathbf{w}$ define a three-dimensional subspace in $\mathbb{R}^n$, given by $v_1 + v_2 + v_3 + 2v_4 - 2 = 0$, where coordinate v_i corresponds to the state of unit u_i. It can be shown that this subspace does not contain other points from $\{0, 1\}^4$. For other points (configurations) the function $g(\mathbf{v}) = v_1 + v_2 + v_3 + 2v_4 - 2$ is either less or greater than 0. Now, by choosing the consensus function $C = -g^2$ we are sure that the consensus is maximal only for the points $\mathbf{x}$, $\mathbf{y}$, $\mathbf{z}$, and $\mathbf{w}$. As g^2 is a quadratic function of the states of the units in the Boltzmann machine, this immediately gives a possible choice of the connections, shown in Figure 10.6. The approach gives a straightforward choice of the connection strengths for a given subset $\mathcal{R}'$, provided the points in $\mathcal{R}'$ determine a subspace that does not contain other points from $\{0, 1\}^n$. However, determining whether other points are contained in the subspace is quite troublesome if the number of units is large. Furthermore, the problem of finding a suitable subset $\mathcal{R}'$ remains open. Nevertheless, the approach seems interesting as it might lead to a better understanding of the inherent difficulty of classification and learning problems [Ten Bosch, 1988].

The problems of finding appropriate internal configurations and corresponding connection strengths can be considered from a combined point of view. Then the internal configurations should be chosen such that the resulting requirements on the connection strengths are not conflicting.

Fortunately, a Boltzmann machine can simultaneously solve these related problems by self-organization. By continuously presenting examples of environmental configurations, the Boltzmann machine adjusts the connection strengths, such that the presented examples eventually correspond to locally maximal configurations of the consensus function. Simultaneously, internal

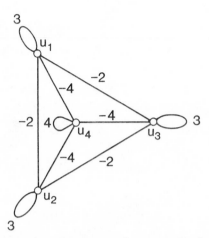

Figure 10.6: A possible choice of the connection strengths, for which the configurations (0001), (0110), (1010), and (1100) have maximal consensus.

representations emerge that capture the regularities in the presented examples.

A detailed discussion of the learning capabilities of the Boltzmann machine is presented in the next chapter.

CHAPTER 11

Learning and Boltzmann Machines

The ability to learn is one of the most essential characteristics of intelligent behaviour. It is therefore not surprising that for many decades learning has attracted the interest of many researchers from such diverse areas as psychology, philosophy, biology, and mathematics. Nevertheless, progress in exposing the basic principles underlying learning processes has been slow and it is fair to say that up to the present it remains a poorly understood subject of interest.

Ever since man started to build machines to improve his living conditions, people have been fascinated by the idea of building machines that learn. The arrival of the digital computer has clearly made these aspirations much more realistic and for many years machine learning has been an active area of research in the field of artificial intelligence[Michalski, Carbonell & Mitchell, 1983]. Artificial systems that learn have been studied from several different perspectives. Interest in these systems is generally motivated by either direct practical use of these systems in for instance industrial situations, or the conviction that studying these artificial learning systems will shed some new light on the fundamental principles of intelligence, human and artificial alike.

The subject of machine learning has been studied from many different perspectives, emphasizing different aspects and goals. Carbonell, Michalski & Mitchell [1983] mention the following three paradigms around which research in machine learning is centred.

1. **Neural Modelling**
   Neural modelling aims at building general purpose learning systems that start with little or no initial structure or task-oriented knowledge. In such systems, learning consists of incremental changes of the probabilities that 'neurons' are activated. As no initial knowledge is assumed, these systems cannot be expected to generate high-level knowledge.

Applications of these systems concentrate on low-level learning in the field of pattern recognition.

2. **Symbolic Concept-Oriented Learning**
   This approach aims at acquiring higher-level knowledge by using logic or graph representations rather than numerical and statistical methods. Systems based on this approach learn symbolic descriptions of high-level concepts by some sort of reasoning process.

3. **Knowledge Intensive Learning**
   Systems based on this approach use task-oriented knowledge and the constraints it provides in guiding the learning process. These systems concentrate on a very specific application area and already possess a great deal of task-specific knowledge.

Here, we consider the learning capabilities of the Boltzmann machine, which clearly follows the first paradigm.

Furthermore, Carbonell, Michalski & Mitchell [1983] give a taxonomy of machine learning research, according to different classification criteria. They give for instance the following classification of machine learning approaches based on the underlying learning strategy.

1. **Rote Learning and Direct Implanting of New Knowledge**
   No inference or transformation of knowledge is required on the part of the learner. Examples of this type of learning are learning by being programmed and learning by memorizing given facts and data.

2. **Learning from Instruction**
   Acquiring knowledge, from a teacher for instance, and requiring that the learner transforms the knowledge to an internally usable representation.

3. **Learning from Analogy**
   Acquiring new facts or skills by transforming and augmenting existing knowledge that bears a strong similarity to the desired new concept or skill.

4. **Learning from Examples**
   Given a set of examples and counterexamples of one or more concepts, the learner induces a general description of these concepts.

5. **Learning from Observation and Discovery**
   This very general form of learning, which is also called unsupervised learning, requires the learner to capture the regularities in its environment and to adjust its internal representation accordingly.

The learning capabilities of a Boltzmann machine can be considered to belong to subclasses 4 and 5. If a Boltzmann machine is used for strict classification purposes (supervised learning) its learning is called *learning from examples*. If a Boltzmann machine is more generally used as an associative or content-addressable memory, its learning is *learning from observation*.

A classification according to the type of knowledge acquired by the learner is also given by Carbonell, Michalski and Mitchell. According to this criterion, learning in a Boltzmann machine can be classified as learning of numerical parameters. For further details and a general overview of machine learning the reader is referred to Michalski, Carbonell & Mitchell [1983].

The learning capabilities of a Boltzmann machine can be considered typical for connectionist network models [Rumelhart, McClelland & the PDP Research Group, 1986].

In this chapter we focus our attention on the ability of a Boltzmann machine to solve classification problems by learning. Later on, we also mention some related pattern recognition problems.

11.1 Learning from Examples

In Chapter 10 we have defined a classification problem as one of automatically generating the corresponding outputs for a number of inputs. An example of a classification problem is that of automatically classifying images of tables and chairs.

To illustrate the full potential of the Boltzmann machine as a learning model, we generalize a classification problem as follows. We not only want a Boltzmann machine to classify given inputs correctly, but also to generate a corresponding input for a given output. Returning to the table-and-chair example, we want the Boltzmann machine to recognize tables and chairs as well as give 'typical' examples of these types of objects.

Given that some of the environmental units in a Boltzmann machine are clamped to a specific state by the environment of the machine, the Boltzmann machine must be able to generate the most probable interpretation of these incoming stimuli, by adjusting the states of the remaining free units. In this way, the Boltzmann machine should be able to obtain the most probable environmental configuration, given that the states of some of the environmental units are fixed.

If the environment clamps environmental configurations (examples) to the environmental units according to a given probability distribution, we want the Boltzmann machine to adjust its connection strengths such that the stationary distribution, defined on the environmental units, becomes identical to the given probability distribution. The more often an example is given to the Boltzmann machine, the higher the corresponding consensus should be. In that case, the Boltzmann machine is able to produce from a partially clamped environmental configuration the complete environmental configuration which is most probable under the given probability distribution. Thus, the Boltzmann machine is able to generate for a given input the most probable output, and for a given output the most probable input.

Learning in a Boltzmann machine can be considered as learning from examples. During the learning process, which is governed by a so-called learning algorithm, examples are presented to the machine as inputs, together with the corresponding classification labels as outputs. For the table-and-chair example, images of tables or chairs are presented as examples. For each image we also 'tell' the Boltzmann machine whether it is an image of a table or a chair.

The learning algorithm described in this chapter is basically due to Hinton & Sejnowski [1983], but it is closely related to the *maximum likelihood methods* used to estimate unknown parameters of exponential distributions [Dempster, Laird & Rubin, 1977] or hidden Markov chains [Bahl, Jelinek & Mercer, 1983]. Maximum likelihood methods iteratively adjust the unknown parameters so as to increase the probability that the generic model will produce the observed data.

Before describing the learning algorithm in greater detail, we first examine in Section 11.2 the equilibrium properties of a Boltzmann machine. These properties play an essential role in the construction of the learning algorithm. In Section 11.3 we describe a learning algorithm for Boltzmann machines without hidden units. In Section 11.4 this learning algorithm is extended to Boltzmann machines with hidden units.

11.2   Equilibrium Properties

In this section we derive some fundamental properties of the Boltzmann machine that are of vital importance to the learning capabilities of the model.

When a Boltzmann machine is in equilibrium, we can distinguish between globally observable behaviour and locally observable behaviour.

Definition 11.1 Let $B = (U, C)$ be a Boltzmann machine, then its *globally observable behaviour*, or *global behaviour* for short, is defined as the stationary distribution $\mathbf{q}(c)$, which gives for all configurations $k \in R$ the probability of occurring at equilibrium; see Theorems 8.1 and 8.2. The stationary distribution is defined by

$$
\begin{aligned}
q_k(c) &= \frac{1}{N_0(c)} \exp\left(\frac{C(k)}{c}\right) \\
&= \frac{1}{N_0(c)} \exp\left(\frac{1}{c} \sum_{\{u,v\} \in C} s_{\{u,v\}} k(u) k(v)\right),
\end{aligned} \tag{11.1}
$$

where $N_0(c) = \sum_{l \in R} \exp(\frac{C(l)}{c})$.

The *locally observable behaviour* or *local behaviour* of a Boltzmann machine is defined as the *activation vector* $\mathbf{p}(c)$, which gives for each connection $\{u, v\} \in C$ the probability of being activated at equilibrium, for some value of the control parameter c. The *activation probability* $p_{\{u,v\}}(c)$ of $\{u, v\}$ is defined as

$$
p_{\{u,v\}}(c) = \sum_{k \in R} q_k(c)\, k(u)\, k(v). \tag{11.2}
$$
∎

Clearly, from (11.2) it follows directly that a global behaviour $\mathbf{q}(c)$ uniquely specifies the corresponding local behaviour $\mathbf{p}(c)$. However, two Boltzmann machines having a different global behaviour might still have the same local behaviour. Below, we will prove that this is not the case, however, by elaborating on the relationship between local and global behaviour of a Boltzmann machine.

Due to a correspondence between its local and global behaviour, a Boltzmann machine is able to learn a desired global behaviour using only locally motivated adjustments of the connection strengths. This is very important, since we want the Boltzmann machine to use its inherent parallelism when learning.

Now, we first derive a one-to-one correspondence between the global behaviour of a Boltzmann machine and the choice of the connection strengths. This correspondence is given in the following theorem

Theorem 11.1 *Let $B = (U, C)$ and $B' = (U', C')$ be isomorphic Boltzmann machines, i.e. $U = U'$ and $C = C'$, with stationary distributions $\mathbf{q}(c)$ and $\mathbf{q}'(c')$,*

respectively. Then $\mathbf{q}(c)$ *and* $\mathbf{q}'(c')$ *are identical if and only if for all* $\{u, v\} \in C$
we have

$$\frac{s_{\{u,v\}}}{c} = \frac{s'_{\{u,v\}}}{c'}. \tag{11.3}$$

Proof From the definition of a stationary distribution it immediately follows that $\mathbf{q}(c)$ and $\mathbf{q}'(c')$ are identical if (11.3) holds for all connections. Thus, we only need to show that two identical stationary distributions yields identical connection strengths for all connections.

If $\mathbf{q}(c) = \mathbf{q}'(c')$ then $q_k(c) = q'_k(c')$ for all configurations $k \in \mathcal{R}$. By making specific choices for k, one can show that (11.3) holds for all $\{u, v\} \in C$.

Firstly, we choose k such that $\sum_{u \in \mathcal{U}} k(u) = 0$. For this choice the corresponding components of the stationary distributions $\mathbf{q}(c)$ and $\mathbf{q}'(c')$ are given by

$$q_k(c) = \frac{1}{N_0(c)} \text{ and } q'_k(c') = \frac{1}{N'_0(c')},$$

respectively. Hence, the normalization constants $N_0(c)$ and $N'_0(c')$ are identical for both Boltzmann machines $\mathcal{B}$ and $\mathcal{B}'$.

Next, we consider all configurations k for which $\sum_{u \in \mathcal{U}} k(u) = 1$. For such a configuration k, where $k(u) = 1$ for exactly one unit u, we have

$$\exp\left(\frac{s_{\{u,u\}}}{c}\right) = \exp\left(\frac{s'_{\{u,u\}}}{c'}\right),$$

if $q_k(c) = q'_k(c')$. Consequently, for all bias connections $\{u, u\} \in C$ we obtain

$$\frac{s_{\{u,u\}}}{c} = \frac{s'_{\{u,u\}}}{c'}. \tag{11.4}$$

Finally, we consider all configurations k for which exactly two units u and v, joined by a connection, are 'on'. From these configurations we can conclude, that for all connections $\{u, v\} \in C$, we have

$$\exp\left(\frac{s_{\{u,u\}} + s_{\{u,v\}} + s_{\{v,v\}}}{c}\right) = \exp\left(\frac{s'_{\{u,u\}} + s'_{\{u,v\}} + s'_{\{v,v\}}}{c'}\right).$$

Using (11.4), we then obtain

$$\frac{s_{\{u,v\}}}{c} = \frac{s'_{\{u,v\}}}{c'},$$

for all $\{u, v\} \in C$, which completes the proof of the theorem. ∎

From Theorem 11.1 we conclude that for a given value of the control parameter c and a given global behaviour $\mathbf{q}(c)$ we can uniquely determine the strengths of the connections. Note that the exact value of the control parameter c is not essential.

If two isomorphic Boltzmann machines $\mathcal{B}$ and $\mathcal{B}'$ have the same global behaviour for different values of their control parameters, i.e. $c \neq c'$, then multiplying the connection strengths of one of the Boltzmann machines, say $\mathcal{B}'$, with a constant factor $\frac{c}{c'}$ results in identical global behaviour of both Boltzmann machines for identical values of their control parameters. Hereafter, we assume identical values of the control parameters, when comparing the equilibrium properties of isomorphic Boltzmann machines.

By assuming that the values of the control parameters are identical for both Boltzmann machines we can derive the following corollary to Theorem 11.1.

Corollary 11.1 *Let $\mathcal{B}$ and $\mathcal{B}'$ be isomorphic Boltzmann machines, then the stationary distributions $\mathbf{q}(c)$ and $\mathbf{q}'(c)$ are identical if and only if $s_{\{u,v\}} = s'_{\{u,v\}}$ for all $\{u, v\} \in C$.* ∎

When learning, a Boltzmann machine adjusts its connection strengths so as to establish some given desired global behaviour. However, not every desired probability distribution $\mathbf{q}(c)$ is feasible. What we mean by a feasible global behaviour is expressed by the following definition.

Definition 11.2 A desired global behaviour $\mathbf{q}(c)$ for a Boltzmann machine $\mathcal{B} = (\mathcal{U}, C)$ is called *feasible* if the strengths of the connections in C can be chosen such that the global behaviour of $\mathcal{B}$ at equilibrium is given by $\mathbf{q}(c)$. ∎

An arbitrary probability distribution may imply contradictory requirements on the connection strengths (cf. Example 10.2), and consequently is infeasible. In general, a global behaviour is feasible if the parameters (connection strengths) can be chosen such that an exponential distribution fits the desired global behaviour. Theorem 11.2 poses a necessary condition on $\mathbf{q}(c)$ for feasibility.

Theorem 11.2 *If global behaviour $\mathbf{q}(c)$ is feasible for some Boltzmann machine $\mathcal{B}$ and some $c \in \mathbb{R}^+$, then for all $k \in \mathcal{R} : 0 < q_k(c) < 1$.*

Proof The theorem is proved by *reductio ad absurdum*. By definition, we have that $0 \leq q_k(c) \leq 1$ for all $k \in \mathcal{R}$. Assume that $\mathbf{q}(c)$ is feasible and that $q_k(c) = 0$ for some configuration k and some $c \in \mathbb{R}^+$. In that case,

$\exp(\frac{C(k)}{c}) = 0$, which would require that $C(k) = -\infty$. However, a finite sum of finite connection strengths necessarily assumes a finite value, which contradicts the feasibility of $\mathbf{q}(c)$. Thus, $q_k(c) > 0$ for all $k \in \mathcal{R}$. Clearly, as $\sum_{k \in \mathcal{R}} q_k(c) = 1$, this implies that $q_k(c) < 1$ for all $k \in \mathcal{R}$. ∎

The condition given in Theorem 11.2 is necessary but not sufficient. This can be shown by an example of a desired global behaviour, which satisfies the condition of Theorem 11.2 but which is nevertheless infeasible. Such a desired global behaviour is given in the following example, which is closely related to Example 10.2.

Example 11.1 Let $\mathcal{B} = (\mathcal{U}, \mathcal{C})$ be a Boltzmann machine where $\mathcal{U} = \{u_1, u_2, u_3\}$ is the set of units and $\mathcal{C} = \{\{u, v\} \mid u, v \in \mathcal{U}\}$ the set of connections. Suppose that the desired global behaviour $\mathbf{q}(c)$ for $c = 1$ is given by Table 11.1. From

k	u_1	u_2	u_3	$q_k(1)$
a	0	0	0	0.20
b	0	0	1	0.05
c	0	1	0	0.05
d	0	1	1	0.20
e	1	0	0	0.05
f	1	0	1	0.20
g	1	1	0	0.20
h	1	1	1	0.05

Table 11.1: For each configuration k the desired probability of occurring at equilibrium is given for $c = 1$.

$q_a(1) = \frac{1}{5}$ and $C(a) = 0$ it follows that $q_k(1) = \frac{1}{5} \exp(C(k))$. Similarly, we can derive from the probabilities $q_e(1)$, $q_c(1)$ and $q_b(1)$ that $s_{\{u_1, u_1\}}$, $s_{\{u_2, u_2\}}$ and $s_{\{u_3, u_3\}}$ are all equal to $- \ln 4$. Furthermore, using the probabilities $q_d(1)$, $q_f(1)$ and $q_g(1)$ we can show that $s_{\{u_1, u_2\}}$, $s_{\{u_1, u_3\}}$ and $s_{\{u_2, u_3\}}$ are all equal to $\ln 16$. In that way, the strengths of all connections in $\mathcal{C}$ are fixed. Using these strengths to calculate $q_h(1)$ yields $q_h(1) = \frac{64}{5}$, which contradicts the requirement that $q_h(1) = \frac{1}{20}$. Hence, the requirements on the connection strengths imposed by $\mathbf{q}(1)$ in Table 11.1 are contradictory. ◼

Similarly to the feasibility of a desired global behaviour we can define the feasibility of a desired local behaviour as follows.

Definition 11.3 A desired local behaviour $\mathbf{p}(c)$ is called *feasible* for a Boltzmann machine $\mathcal{B} = (\mathcal{U}, \mathcal{C})$ if the strengths of the connections in $\mathcal{C}$ can be chosen such that the local behaviour of $\mathcal{B}$ at equilibrium is given by $\mathbf{p}(c)$. ◼

Theorem 11.3 states some necessary conditions for feasibility of a desired local behaviour $\mathbf{p}(c)$. These conditions again can be shown not to be sufficient.

Theorem 11.3 *If a desired local behaviour* $\mathbf{p}(c)$ *is feasible for some Boltzmann machine* $\mathcal{B} = (\mathcal{U}, \mathcal{C})$, *then for all connections* $\{u, v\} \in \mathcal{C}$ *we have*

$$0 < p_{\{u,v\}}(c) < p_{\{u,u\}}(c) < 1 \tag{11.5}$$

and

$$p_{\{u,u\}}(c) + p_{\{v,v\}}(c) - p_{\{u,v\}}(c) < 1. \tag{11.6}$$

Proof Considering the definition of $\mathbf{p}(c)$ and taking into account the necessary feasibility condition of Theorem 11.2 it is easy to see that (11.5) is a necessary condition. Furthermore, the probability P that none of the connections $\{u, u\}$, $\{u, v\}$ and $\{v, v\}$ are activated is given by

$$P = 1 - \left[p_{\{u,u\}}(c) + p_{\{v,v\}}(c) - p_{\{u,v\}}(c) \right].$$

As each configuration has a positive probability of occurring at equilibrium, P must be positive. This yields the necessity of (11.6). ◼

Given the definition of $\mathbf{p}(c)$, it is easy to show that *global equality*, i.e. equality of the global behaviour of $\mathcal{B}$ and $\mathcal{B}'$, induces *local equality*, i.e. equality of the local behaviour of $\mathcal{B}$ and $\mathcal{B}'$. Below, we show in a number of steps that local equality also induces global equality. To that end, we first introduce a measure $D(\mathbf{q}|\mathbf{q}')$, called the divergence, which indicates how much the global behaviour $\mathbf{q}(c)$ of a Boltzmann machine $\mathcal{B} = (\mathcal{U}, \mathcal{C})$ differs from the given global behaviour $\mathbf{q}'(c)$ of an isomorphic Boltzmann machine $\mathcal{B}'$. The value of $D(\mathbf{q}|\mathbf{q}')$ clearly depends on the choice of the strengths of the connections in $\mathcal{C}$, since global behaviour $\mathbf{q}(c)$ strongly depends on the choice of the connection strengths. Obviously, $D(\mathbf{q}|\mathbf{q}')$ is minimal if the strengths of the connections in $\mathcal{B}$ are chosen identical to the strengths of the connections in $\mathcal{B}'$. By representing a specific choice of the connection strengths as a vector $\mathbf{s} \in \mathbb{R}^{|\mathcal{C}|}$, such that each component gives the strength of a connection in $\mathcal{C}$, we consider $D(\mathbf{q}|\mathbf{q}')$ as an $|\mathcal{C}|$-dimensional function, assigning to each choice **s** of the connection strengths a value indicating to what extent the corresponding global behaviour $\mathbf{q}(c)$ of $\mathcal{B}$ differs from the given global behaviour $\mathbf{q}'(c)$ of $\mathcal{B}'$.

Definition 11.4 Let $\mathcal{B}$ and $\mathcal{B}'$ be isomorphic Boltzmann machines with global behaviour $\mathbf{q}(c)$ and $\mathbf{q}'(c)$, respectively. Then the *divergence* $D(\mathbf{q}|\mathbf{q}')$ of $\mathbf{q}(c)$ with respect to $\mathbf{q}'(c)$ is given by

$$D(\mathbf{q}|\mathbf{q}') = \sum_{k \in \mathcal{R}} q'_k(c) \ln \frac{q'_k(c)}{q_k(c)}. \tag{11.7}$$

∎

This divergence measure is well known in information theory. There it serves as a measure of information in order to discriminate in favour of one of two hypotheses [Kullback, 1959]. Note that $D(\mathbf{q}|\mathbf{q}')$ does not satisfy all requirements to be a metric [Simmons, 1963], as $D(\mathbf{q}|\mathbf{q}')$ is not symmetric: the divergence of $\mathbf{q}(c)$ with respect to $\mathbf{q}'(c)$ is usually not identical to the divergence of $\mathbf{q}'(c)$ with respect to $\mathbf{q}(c)$. However, $D(\mathbf{q}|\mathbf{q}')$ satisfies the following important requirement.

Lemma 11.1 *Let $\mathcal{B}$ and $\mathcal{B}'$ be isomorphic Boltzmann machines with global behaviour $\mathbf{q}(c)$ and $\mathbf{q}'(c)$, respectively, and let $c > 0$, then the divergence $D(\mathbf{q}|\mathbf{q}') = 0$ if and only if $\mathbf{q}(c) = \mathbf{q}'(c)$, and $D(\mathbf{q}|\mathbf{q}') > 0$ otherwise.*

Proof By examining the function $f(x) = \ln x + \frac{1}{x} - 1$ and its derivative $f'(x)$ it is straightforward to show that for $x = 1$ we have

$$\ln x = 1 - \frac{1}{x},$$

and that for all other $x \in \mathbb{R}^+$ we have

$$\ln x > 1 - \frac{1}{x}.$$

Consequently,

$$\begin{aligned}
D(\mathbf{q}|\mathbf{q}') &= \sum_{k \in \mathcal{R}} q'_k(c) \ln \frac{q'_k(c)}{q_k(c)} \\
&\geq \sum_{k \in \mathcal{R}} q'_k(c) \left(1 - \frac{q_k(c)}{q'_k(c)} \right) \\
&= \sum_{k \in \mathcal{R}} [q'_k(c) - q_k(c)]
\end{aligned}$$

$$= \sum_{k \in \mathcal{R}} q'_k(c) - \sum_{k \in \mathcal{R}} q_k(c)$$

$$= 0.$$

Furthermore, since $\ln x = 1 - \frac{1}{x}$ if and only if $x = 1$, it immediately follows that $D(\mathbf{q}|\mathbf{q}') = 0$ if and only if $q'_k(c) = q_k(c)$ for all $k \in \mathcal{R}$. ∎

By examining the properties of the divergence $D(\mathbf{q}|\mathbf{q}')$ we next prove that local equality induces global equality. From Lemma 11.1 and Corollary 11.1 we know that $D(\mathbf{q}|\mathbf{q}')$ assumes its minimal value in exactly one point $\mathbf{s}_{min} \in \mathbb{R}^{|C|}$, for which $\mathbf{q}(c) = \mathbf{q}'(c)$. For $\mathbf{s}_{min}$ the divergence is equal to 0. Since global equality induces local equality, also $\mathbf{p}(c) = \mathbf{p}'(c)$ for $\mathbf{s}_{min}$.

Next, we prove that $\mathbf{s}_{min}$ is the only $\mathbf{s} \in \mathbb{R}^{|C|}$ for which $\mathbf{p}(c) = \mathbf{p}'(c)$. This is shown by using the following lemmas.

Lemma 11.2 *Let $\mathcal{B}$ and $\mathcal{B}'$ be isomorphic Boltzmann machines with global behaviour $\mathbf{q}(c)$ and $\mathbf{q}'(c)$, respectively, and local behaviour $\mathbf{p}(c)$ and $\mathbf{p}'(c)$, respectively, then we have*

$$\nabla D(\mathbf{q}|\mathbf{q}') = 0 \quad \Leftrightarrow \quad \mathbf{p}(c) = \mathbf{p}'(c). \tag{11.8}$$

Proof By definition, the gradient $\nabla D(\mathbf{q}|\mathbf{q}') = 0$ if and only if the partial derivative of $D(\mathbf{q}|\mathbf{q}')$ with respect to $s_{\{u,v\}}$ equals 0 for all $\{u, v\} \in C$. The partial derivative of $D(\mathbf{q}|\mathbf{q}')$ with respect to a connection strength $s_{\{u,v\}}$ is given by

$$\frac{\partial D(\mathbf{q}|\mathbf{q}')}{\partial s_{\{u,v\}}} = -\sum_{k \in \mathcal{R}} \frac{q'_k(c)}{q_k(c)} \frac{\partial q_k(c)}{\partial s_{\{u,v\}}}.$$

The partial derivative of $q_k(c)$ with respect to $s_{\{u,v\}}$ is given by

$$\frac{\partial q_k(c)}{\partial s_{\{u,v\}}} = \frac{k(u)\, k(v)\, \exp(\frac{C(k)}{c})}{c N_0(c)} - \frac{\exp(\frac{C(k)}{c})\, p_{\{u,v\}}(c)}{c N_0(c)}$$

$$= \frac{k(u)\, k(v)\, q_k(c)}{c} - \frac{q_k(c)\, p_{\{u,v\}}(c)}{c}$$

$$= \frac{q_k(c)}{c} \left[k(u)\, k(v) - p_{\{u,v\}}(c) \right], \tag{11.9}$$

where $p_{\{u,v\}}(c)$ is defined as in (11.2). Using (11.9), we then obtain

$$\frac{\partial D(\mathbf{q}|\mathbf{q}')}{\partial s_{\{u,v\}}} = -\frac{1}{c} \sum_{k \in \mathcal{R}} \frac{q'_k(c)}{q_k(c)} q_k(c) \left[k(u)\, k(v) - p_{\{u,v\}}(c) \right]$$

$$= -\frac{1}{c} \sum_{k \in \mathcal{R}} q'_k(c) \left[k(u)\, k(v) - p_{\{u,v\}}(c) \right].$$

Since $\sum\limits_{k \in \mathcal{R}} q'_k(c) \, k(u) \, k(v) = p'_{\{u,v\}}(c)$ and $\sum\limits_{k \in \mathcal{R}} q'_k(c) = 1$, we finally obtain

$$\frac{\partial D(\mathbf{q}|\mathbf{q}')}{\partial s_{\{u,v\}}} = -\frac{1}{c}\left[p'_{\{u,v\}}(c) - p_{\{u,v\}}(c)\right]. \tag{11.10}$$

Thus, $\nabla D(\mathbf{q}|\mathbf{q}') = 0$ if and only if $\mathbf{p}(c) = \mathbf{p}'(c)$.   ∎

Lemma 11.3 *The divergence $D(\mathbf{q}|\mathbf{q}')$ of the global behaviour $\mathbf{q}(c)$ of a Boltzmann machine $B = (\mathcal{U}, \mathcal{C})$ with respect to the global behaviour $\mathbf{q}'(c)$ of a given isomorphic Boltzmann machine B' is a strictly convex function in $\mathbb{R}^{|\mathcal{C}|}$.*

Proof We first prove that $D(\mathbf{q}|\mathbf{q}')$ is convex, which it is if the Hessian matrix H is positive semidefinite, i.e. if $\mathbf{s}^T H \mathbf{s} \geq 0$ for every $\mathbf{s} \in \mathbb{R}^{|\mathcal{C}|}$ [Luenberger, 1973], where a component of the Hessian matrix H is given by

$$\frac{\partial^2 D(\mathbf{q}|\mathbf{q}')}{\partial s_{\{w,z\}} \partial s_{\{u,v\}}}. \tag{11.11}$$

For reasons of convenience, we introduce a stochastic variable $\mathbf{Z}_{\{u,v\}}$ for each connection $\{u, v\}$, defined by

$$\mathbf{Z}_{\{u,v\}} = \begin{cases} 1 & \text{if } \{u, v\} \text{ is activated} \\ 0 & \text{if } \{u, v\} \text{ is not activated.} \end{cases}$$

Clearly, at equilibrium, $\mathbb{E}_c(\mathbf{Z}_{\{u,v\}})$ equals the activation probability $p_{\{u,v\}}(c)$ of connection $\{u, v\}$. Using (11.10), the second partial derivative of $D(\mathbf{q}|\mathbf{q}')$ is given by

$$\frac{\partial^2 D(\mathbf{q}|\mathbf{q}')}{\partial s_{\{w,z\}} \partial s_{\{u,v\}}} = -\frac{1}{c}\frac{\partial}{\partial s_{\{w,z\}}}[p'_{\{u,v\}}(c) - p_{\{u,v\}}(c)].$$

The local behaviour of B' clearly does not depend on the choice of connection strengths of B. Consequently,

$$\begin{aligned}
\frac{\partial^2 D(\mathbf{q}|\mathbf{q}')}{\partial s_{\{w,z\}} \partial s_{\{u,v\}}} &= \frac{1}{c}\frac{\partial p_{\{u,v\}}(c)}{\partial s_{\{w,z\}}} \\
&= \frac{1}{c}\frac{\partial}{\partial s_{\{w,z\}}}\sum_{k \in \mathcal{R}} k(u) \, k(v) \, q_k(c) \\
&= \frac{1}{c^2}\left[p_{\{u,v,w,z\}}(c) - p_{\{u,v\}}(c) \, p_{\{w,z\}}(c)\right] \\
&= \frac{1}{c^2}\left[\mathbb{E}_c(\mathbf{Z}_{\{u,v\}}\mathbf{Z}_{\{w,z\}}) - \mathbb{E}_c(\mathbf{Z}_{\{u,v\}}) \, \mathbb{E}_c(\mathbf{Z}_{\{w,z\}})\right] \\
&= \frac{1}{c^2} \operatorname{Cov}_c(\mathbf{Z}_{\{u,v\}}, \mathbf{Z}_{\{w,z\}}),
\end{aligned}$$

where

$$p_{\{u,v,w,z\}}(c) = \sum_{k \in \mathcal{R}} k(u) \, k(v) \, k(w) \, k(z) \, q_k(c)$$

denotes the probability that units u, v, w, and z are all 'on' at the same time, if the Boltzmann machine has reached equilibrium, and $\mathrm{Cov}_c(\mathbf{Z}_{\{u,v\}}, \mathbf{Z}_{\{w,z\}})$ denotes the *covariance* of $\mathbf{Z}_{\{u,v\}}$ and $\mathbf{Z}_{\{w,z\}}$. Hence, the second partial derivative of (11.11) is proportional to the covariance of the stochastic variables $\mathbf{Z}_{\{u,v\}}$ and $\mathbf{Z}_{\{w,z\}}$, and the Hessian matrix H equals the *covariance matrix*. Thus, for all vectors $\mathbf{s} \in \mathbb{R}^{|\mathcal{C}|}$ we obtain

$$
\begin{aligned}
\mathbf{s}^T H \mathbf{s} &= \frac{1}{c^2} \sum_{\{u,v\} \in \mathcal{C}} \sum_{\{w,z\} \in \mathcal{C}} \mathrm{Cov}_c(\mathbf{Z}_{\{u,v\}}, \mathbf{Z}_{\{w,z\}}) s_{\{u,v\}} \, s_{\{w,z\}} \\
&= \frac{1}{c^2} \mathbb{E}_c \left(\left(\sum_{\{u,v\} \in \mathcal{C}} s_{\{u,v\}} \mathbf{Z}_{\{u,v\}} \right) \left(\sum_{\{w,z\} \in \mathcal{C}} s_{\{w,z\}} \mathbf{Z}_{\{w,z\}} \right) \right) \\
&= \frac{1}{c^2} \mathbb{E}_c \left(\left(\sum_{\{u,v\} \in \mathcal{C}} s_{\{u,v\}} \mathbf{Z}_{\{u,v\}} \right)^2 \right) \geq 0,
\end{aligned}
\tag{11.12}
$$

which proves that $D(\mathbf{q}|\mathbf{q}')$ is convex. Moreover, since each configuration in a Boltzmann machine has a positive probability of occurring at equilibrium,

$$\mathbb{E}_c \left(\left(\sum_{\{u,v\} \in \mathcal{C}} s_{\{u,v\}} \mathbf{Z}_{\{u,v\}} \right)^2 \right) > 0, \quad \text{for all } \mathbf{s} \neq \mathbf{0},$$

which proves that $D(\mathbf{q}|\mathbf{q}')$ is strictly convex. ∎

We can now prove Theorem 11.4, stating a strict relationship between local equality and global equality.

Theorem 11.4 *Let $\mathcal{B}$ and $\mathcal{B}'$ be isomorphic Boltzmann machines with a global behaviour $\mathbf{q}(c)$ and $\mathbf{q}'(c)$, respectively, and a local behaviour $\mathbf{p}(c)$ and $\mathbf{p}'(c)$, respectively. The relationship between the global behaviour and local behaviour is expressed by*

$$\mathbf{q}(c) = \mathbf{q}'(c) \iff \mathbf{p}(c) = \mathbf{p}'(c). \tag{11.13}$$

Proof As already mentioned, it is easy to show that global equality induces local equality. This follows immediately from the definition of $\mathbf{p}(c)$. Combining the results of Lemmas 11.1, 11.2, and 11.3, we can now also prove that local equality induces global equality. From Lemma 11.3 we conclude that $\nabla D(\mathbf{q}|\mathbf{q}') = 0$ for exactly one $\mathbf{s}$ in $\mathbb{R}^{|\mathcal{C}|}$. Using Lemma 11.2, we obtain that $\mathbf{p}(c) = \mathbf{p}'(c)$ for exactly one $\mathbf{s}$ in $\mathbb{R}^{|\mathcal{C}|}$. From Lemma 11.1 we already concluded that $\mathbf{p}(c) = \mathbf{p}'(c)$ for the global minimum $\mathbf{s}_{min}$, for which $\mathbf{q}(c) = \mathbf{q}'(c)$. Thus, if $\mathbf{p}(c) = \mathbf{p}'(c)$ then also $\mathbf{q}(c) = \mathbf{q}'(c)$, which completes the proof of the theorem. ∎

Hence, a one-to-one correspondence exists between the local and global behaviour of a Boltzmann machine. This fundamental property can be used to obtain a desired (feasible) global behaviour, by adjusting the strengths of the connections on the basis of their activation probabilities only.

We end this section with a corollary that can be straightforwardly proved by combining the results of Corollary 11.1 and Theorem 11.4.

Corollary 11.2 *Let $\mathcal{B}$ and $\mathcal{B}'$ be two isomorphic Boltzmann machines where $\mathcal{U}$ is the set of units and $\mathcal{C}$ the set of connections, then we have*

$$\forall \{u, v\} \in \mathcal{C} : \ s_{\{u,v\}} = s'_{\{u,v\}}$$

$$\Leftrightarrow$$

$$\forall \{u, v\} \in \mathcal{C} : \ p_{\{u,v\}}(c) = p'_{\{u,v\}}(c). \tag{11.14}$$

∎

Corollary 11.2 implies that a given local behaviour of a Boltzmann machine uniquely determines the strengths of its connections.

11.3 Learning without Hidden Units

In this section we show that a Boltzmann machine $\mathcal{B}$ is able to learn a given (feasible) global behaviour $\mathbf{q}'(c)$, just by seeing examples. During the learning process, the strengths of the connections are iteratively adjusted such that the global behaviour $\mathbf{q}(c)$ of $\mathcal{B}$ converges to the required global behaviour $\mathbf{q}'(c)$.

To effectively exploit the inherent parallelism of a Boltzmann machine we restrict ourselves to learning algorithms that only use locally available information for adjusting the connection strengths. Using the equilibrium properties, discussed in the previous section, we can now formulate the learning algorithm for Boltzmann machines without hidden units.

The goal of the learning algorithm can be formulated as follows: *Given a desired global behaviour $\mathbf{q}'(c)$, minimize the divergence $D(\mathbf{q}|\mathbf{q}')$ of the global behaviour $\mathbf{q}(c)$ of $\mathcal{B}$ with respect to the given behaviour $\mathbf{q}'(c)$.*

The divergence $D(\mathbf{q}|\mathbf{q}')$ is a strictly convex function, which implies that an iterative improvement algorithm that minimizes $D(\mathbf{q}|\mathbf{q}')$ cannot get stuck in poor local minima. From Lemma 11.2 we know that the partial derivative of $D(\mathbf{q}|\mathbf{q}')$ with respect to a connection strength $s_{\{u,v\}}$ is given by

$$\frac{\partial D(\mathbf{q}|\mathbf{q}')}{\partial s_{\{u,v\}}} = -\frac{1}{c}\left[p'_{\{u,v\}}(c) - p_{\{u,v\}}(c) \right], \tag{11.15}$$

where $p'_{\{u,v\}}(c)$ denotes the activation probability of $\{u,v\}$ when the global behaviour of $\mathcal{B}$ is given by the desired behaviour $\mathbf{q}'(c)$, and $p_{\{u,v\}}(c)$ denotes the activation probability of $\{u,v\}$ for $\mathbf{s}$, the present choice of the connection strengths.

Note that we only need to know the activation probabilities $p'_{\{u,v\}}(c)$ and $p_{\{u,v\}}(c)$ of the corresponding connection to calculate this partial derivative. These activation probabilities can be estimated by using only locally available information.

11.3.1 Outline of the Learning Algorithm

Before explaining how the activation probabilities can be estimated, we give an outline of the learning algorithm. The algorithm starts off with an arbitrary choice $\mathbf{s}^{(0)}$ of the connection strengths, e.g. $\mathbf{s}^{(0)} = \mathbf{0}$, that is all connections are given an initial strength equal to zero. Next, a sequence $\{\mathbf{s}^{(k)}\}$, $k = 1, 2, 3, \ldots$, is iteratively constructed by

$$\mathbf{s}^{(i+1)} = \mathbf{s}^{(i)} - \beta \nabla D(\mathbf{q}^{(i)}|\mathbf{q}'), \tag{11.16}$$

where $\mathbf{q}^{(i)}$ denotes the global behaviour corresponding to $\mathbf{s}^{(i)}$, and $\beta \in \mathbf{R}^+$ some constant. In the $(i+1)^{th}$ iteration, the next point $\mathbf{s}^{(i+1)} \in \mathbf{R}^{|\mathcal{C}|}$ is chosen by moving from $\mathbf{s}^{(i)}$ in the direction opposite to the gradient $\nabla D(\mathbf{q}^{(i)}|\mathbf{q}')$, i.e. in the direction of the steepest descent. The step length is chosen proportional to the gradient in $\mathbf{s}^{(i)}$. The coefficient β can be chosen such that the method is *stable*.

Definition 11.5 Let f be a function defined in $\mathbf{R}^{|\mathcal{C}|}$, then a minimization algorithm that iteratively generates a sequence $\{\mathbf{s}^{(k)}\}$, $k = 1, 2, 3, \ldots$, is called *stable* if for all $i \in \mathbf{N}$

$$f(\mathbf{s}^{(i+1)}) \leq f(\mathbf{s}^{(i)}). \tag{11.17}$$

∎

In Theorem 11.5 we give an upper bound on β that guarantees the stability of the learning algorithm. To prove this theorem we first give the following lemma, stating an upper bound on the maximum curvature of $D(\mathbf{q}|\mathbf{q}')$.

Lemma 11.4 *Let $D(\mathbf{q}|\mathbf{q}')$ be the divergence of the global behaviour $\mathbf{q}(c)$ of a Boltzmann machine $\mathcal{B} = (\mathcal{U}, \mathcal{C})$ with respect to a desired global behaviour $\mathbf{q}'(c)$. Then the maximum curvature of $D(\mathbf{q}|\mathbf{q}')$ is bounded by $\frac{|\mathcal{C}|}{c^2}$.*

Proof From (11.12) we know that for all $\mathbf{s} \in \mathbb{R}^{|\mathcal{C}|}$

$$\mathbf{s}^T H \mathbf{s} = \frac{1}{c^2} \, \mathbb{E}_c \left(\left(\sum_{\{u,v\} \in \mathcal{C}} s_{\{u,v\}} Z_{\{u,v\}} \right)^2 \right).$$

As $Z_{\{u,v\}}$ equals 0 or 1, for all $\{u, v\} \in \mathcal{C}$, we obtain

$$\mathbf{s}^T H \mathbf{s} \leq \frac{1}{c^2} \, \mathbb{E}_c \left(\left(\sum_{\{u,v\} \in \mathcal{C}} |s_{\{u,v\}}| \right)^2 \right).$$

Assuming $||\mathbf{s}|| = 1$, and using that

$$\sum_{\{u,v\} \in \mathcal{C}} s_{\{u,v\}} \leq \sqrt{|\mathcal{C}|},$$

for each vector $\mathbf{s} \in \mathbb{R}^{|\mathcal{C}|}$ with $||\mathbf{s}|| = 1$, we have

$$\mathbf{s}^T H \mathbf{s} \leq \frac{1}{c^2} \, \mathbb{E}_c \left(\left(\sqrt{|\mathcal{C}|} \right)^2 \right) = \frac{|\mathcal{C}|}{c^2},$$

which gives an upper bound on the maximum curvature of $D(\mathbf{q}|\mathbf{q}')$ of $\frac{|\mathcal{C}|}{c^2}$. ∎

Using Lemma 11.4 we can now prove the following theorem.

Theorem 11.5 *Let $D(\mathbf{q}|\mathbf{q}')$ be the divergence of the global behaviour $\mathbf{q}(c)$ of a Boltzmann machine $\mathcal{B} = (\mathcal{U}, \mathcal{C})$ with respect to a desired global behaviour $\mathbf{q}'(c)$. Then the iterative method for minimizing $D(\mathbf{q}|\mathbf{q}')$, defined by*

$$\mathbf{s}^{(i+1)} = \mathbf{s}^{(i)} - \beta \nabla D(\mathbf{q}^{(i)}|\mathbf{q}'), \tag{11.18}$$

is stable if $\beta \leq \frac{c^2}{|\mathcal{C}|}$.

Proof If we move from $\mathbf{s}^{(i)}$ to $\mathbf{s}^{(i+1)}$ in the opposite direction of the gradient, we are sure to go downhill at first. If $\mathbf{s}^{(i+1)}$ is chosen close enough to

$\mathbf{s}^{(i)}$ we can prove that $D(\mathbf{q}^{(i+1)}|\mathbf{q}') < D(\mathbf{q}^{(i)}|\mathbf{q}')$. Starting with a slope given by $-\nabla D(\mathbf{q}^{(i)}|\mathbf{q}')$ we move over a distance a in the opposite direction of the gradient. The upper bound on the maximum curvature guarantees that we continue to go downhill if $a \leq \frac{|C|}{c^2}\nabla D(\mathbf{q}^{(i)}|\mathbf{q}')$. This is the case if the value of β in (11.18) is chosen such that $\beta \leq \frac{c^2}{|C|}$, which completes the proof of Theorem 11.6. ∎

Consequently, as $D(\mathbf{q}|\mathbf{q}')$ is strictly convex, the sequence $\{\mathbf{s}^{(k)}\}$ generated by the iterative method converges to $\mathbf{s}_{min}$, the global minimum of $D(\mathbf{q}|\mathbf{q}')$, for which $\mathbf{q}(c) = \mathbf{q}'(c)$.

The iterative method that minimizes $D(\mathbf{q}|\mathbf{q}')$ is a simple gradient method whose convergence properties are not as good as those of the well-known steepest descent method [Luenberger, 1973]. This method also moves in the direction of the steepest descent, according to

$$\mathbf{s}^{(i+1)} = \mathbf{s}^{(i)} - \beta^{(i)}\nabla D(\mathbf{q}^{(i)}|\mathbf{q}'). \tag{11.19}$$

The coefficient $\beta^{(i)}$ is here chosen such that it minimizes $D(\mathbf{q}|\mathbf{q}')$ on the line given by the steepest descent.

Note that using the steepest descent method requires the use of non-local information to calculate the point where $D(\mathbf{q}|\mathbf{q}')$ assumes its minimal value. Clearly, this violates the principle of locality.

11.3.2 Estimation of Activation Probabilities

In this section we discuss how the activation probabilities can be estimated by using local information only. The activation probabilities $p'_{\{u,v\}}(c)$ and $p_{\{u,v\}}(c)$ can be estimated as follows. By presenting a number of examples to the Boltzmann machine, the activation probability $p'_{\{u,v\}}(c)$ can be estimated in parallel for all connections in C. Configurations are clamped to the units with a probability given by the desired probability distribution $\mathbf{q}'(c)$. During the presentation of the examples, each connection $\{u, v\}$ counts the number of times it is activated. From this number the relative activation frequency $z'_{\{u,v\}}(c)$ is determined. The relative activation frequency $z'_{\{u,v\}}(c)$ is an approximation of the actication probability $p'_{\{u,v\}}(c)$

The activation probabilities $p_{\{u,v\}}(c)$ are estimated for the the connection strengths given by $\mathbf{s}^{(i)}$, by equilibrating the system at a given value of the control parameter c. Here, equilibration refers to establishing an equilibrium distribution on the units that can freely adjust their states. After equilibration, the activation probabilities $p_{\{u,v\}}(c)$ are approximated for all connections in parallel, by determining the relative activation frequency $z_{\{u,v\}}(c)$ for each $\{u, v\} \in C$.

Hence, each iteration of the learning algorithm can be subdivided into two phases, viz. a *clamped phase* and a *free phase*. In the clamped phase, the Boltzmann machine monitors its environment for a period of time to calculate the relative activation frequencies $z'_{\{u,v\}}(c)$, which are estimates of the activation probabilities $p'_{\{u,v\}}(c)$ that correspond to the desired global behaviour. In the free phase, the Boltzmann machine equilibrates and monitors its own behaviour at equilibrium to calculate the relative activation frequencies $z_{\{u,v\}}(c)$, which are estimates of the activation probabilities $p_{\{u,v\}}(c)$ that correspond to the present global behaviour. At the end of each iteration the connections are adjusted according to (11.16), based on the approximations $z'_{\{u,v\}}(c)$ and $z_{\{u,v\}}(c)$ The learning algorithm is given as a pseudo PASCAL program in Figure 11.1.

---

procedure LEARNING_ ALGORITHM;

begin

    for all $\{u, v\} \in C$ **do**

    $s_{\{u,v\}} := 0;$

    repeat

        CLAMPED_ PHASE;

        FREE_ PHASE;

        $s_{\{u,v\}} := s_{\{u,v\}} - \frac{1}{c} \left[z_{\{u,v\}}(c) - z'_{\{u,v\}}(c) \right] ;$

    until stopcriterion;

  end;

Figure 11.1: A pseudo PASCAL program of the learning algorithm.

The learning algorithm simply does the following. In Definition 8.3, the strength of a connection $\{u, v\}$ is defined as a measure of the desirability that the connection is activated. If a connection is activated less often than required by the desired global behaviour, its strength is increased. If a connection is activated more often than required, its strength is decreased.

Note that in order to prove that $D(\mathbf{q}|\mathbf{q}')$ is strictly convex, $\mathbf{q}'(c)$ was not required to be feasible. Consequently, regardless of whether the desired global behaviour is feasible, the learning algorithm always converges to a unique minimum s_{min} of $D(\mathbf{q}|\mathbf{q}')$. Clearly, we have $\nabla D(\mathbf{q}|\mathbf{q}') = 0$ in s_{min}, which implies that, after learning, the local behaviour $\mathbf{p}(c)$ of a Boltzmann machine is the same as the local behaviour $\mathbf{p}'(c)$ that corresponds to the desired global behaviour $\mathbf{q}'(c)$. Thus, each local behaviour $\mathbf{p}'(c)$ that can be associated with a global behaviour $\mathbf{q}'(c)$, regardless of whether $\mathbf{q}'(c)$ is feasible or not, can be realized in a Boltzmann machine, and is consequently feasible.

Summarizing, we have shown the following learning behaviour of the Boltzmann machine. If, during learning, the examples are clamped to the units of a Boltzmann machine, according to a given global behaviour $\mathbf{q}'(c)$, then the Boltzmann machine is able to adjust its connection strengths such that the resulting global behaviour $\mathbf{q}(c)$ diverges minimally from $\mathbf{q}'(c)$. If $\mathbf{q}'(c)$ is feasible, then $\mathbf{q}(c) = \mathbf{q}'(c)$. If $\mathbf{q}'(c)$ is infeasible, then the Boltzmann machine converges to the feasible global behaviour $\mathbf{q}(c)$ that diverges minimally from $\mathbf{q}'(c)$. Furthermore, the local behaviour $\mathbf{p}(c)$ of the Boltzmann machine equals the local behaviour $\mathbf{p}'(c)$ which corresponds to $\mathbf{q}'(c)$.

We end this section by revisiting the learning problem described in Example 11.1.

Example 11.2 Let us again consider the desired global behaviour $\mathbf{q}'(1)$ specified in Table 11.1. In Example 11.1 we showed that this desired global behaviour is infeasible. If the Boltzmann machine nevertheless tries to learn $\mathbf{q}'(1)$, then it converges to a choice of the connection strengths s_{min}, such that its local behaviour $\mathbf{p}(1)$ equals the local behaviour $\mathbf{p}'(1)$ associated with the desired global behaviour $\mathbf{q}'(1)$. The desired local behaviour $\mathbf{p}'(1)$ can be derived straightforwardly from Table 11.1, and is given by

$$\mathbf{p}'_{\{u,v\}}(c) = \begin{cases} 0.50 & \text{if } u = v \\ 0.25 & \text{if } u \neq v, \end{cases} \tag{11.20}$$

where $u, v \in \{u_1, u_2, u_3\}$. The unique global behaviour $\mathbf{q}(1)$, which corresponds to this local behaviour and minimally diverges from $\mathbf{q}'(1)$, is given by $q_k(1) = 0.125$, for all $k \in \mathcal{R}$. This behaviour is established if $s_{\{u,v\}} = 0$, for all $\{u, v\} \in \mathcal{C}$. Note that the feasible global behaviour $\mathbf{q}(1)$ and the infeasible global behaviour $\mathbf{q}'(1)$ both correspond to the same local behaviour. The divergence $D(\mathbf{q}|\mathbf{q}')$ of $\mathbf{q}(1)$ with respect to $\mathbf{q}'(1)$ is then given by $D(0) = 0.8 \ln 1.6 + 0.2 \ln 0.4$. ∎

11.4 Learning with Hidden Units

In the previous section we discussed a learning algorithm for Boltzmann machines consisting of environmental units only. In this section we show that this algorithm can be extended to Boltzmann machines with hidden units. In Example 10.2 we have seen a classification problem for which hidden units are necessary to classify the given inputs correctly. It can be shown that for such a problem the corresponding desired global behaviour is infeasible if no hidden units are used.

For many problems, the use of hidden units is necessary to obtain a desired global behaviour. Hidden units allow a Boltzmann machine to capture the regularities in its environment that cannot simply be expressed as co-occurrences of the states of pairs of environmental units. The states of hidden units are never directly determined by the environment, but through a process of learning the Boltzmann machine eventually develops a correspondence between environmental configurations - clamped as examples to the environmental units - and internal configurations - which are determined by the states of the hidden units - in such a way that the resulting global behaviour corresponds to the desired behaviour of the environmental units. For that purpose, the Boltzmann machine has to construct an internal representation which captures the regularities in the desired behaviour of the environmental units.

To extend the learning algorithm of the previous section to Boltzmann machines with hidden units, we adjust the definitions given in Sections 11.2 and 11.3 for Boltzmann machines with hidden units.

Definition 11.6 A *visible* global behaviour $\mathbf{v}(c)$ of a Boltzmann machine $\mathcal{B}$ gives, for each environmental configuration $l \in \mathcal{V}$, the probability $v_l(c)$ that the Boltzmann machine is in environmental configuration l, under the assumption that the Boltzmann machine has reached equilibrium. The visible behaviour $\mathbf{v}(c)$ is defined by

$$v_l(c) = \sum_{k \in \mathcal{R}_{\mathcal{V}(l)}} q_k(c), \qquad (11.21)$$

where $\mathcal{R}_{\mathcal{V}(l)} = \{k \in \mathcal{R} \mid \forall v \in \mathcal{U}_{io} : k(v) = l(v)\}$. Now, we define the divergence $D(\mathbf{v}|\mathbf{v}')$ of visible behaviour $\mathbf{v}(c)$ of a Boltzmann machine $\mathcal{B}$ with respect to a desired visible behaviour $\mathbf{v}'(c)$ as follows:

$$D(\mathbf{v}|\mathbf{v}') = \sum_{l \in \mathcal{V}} v_l'(c) \ln \frac{v_l'(c)}{v_l(c)}. \qquad (11.22)$$

∎

The goal of the learning algorithm can now be formulated as: *Minimize the divergence $D(\mathbf{v}|\mathbf{v}')$ by adjusting the connection strengths so that the visible behaviour $\mathbf{v}(c)$ of $\mathcal{B}$ is eventually the same as the desired visible behaviour $\mathbf{v}'(c)$.*

Below, we show that the learning algorithm, developed for Boltzmann machines without hidden units, can be extended to Boltzmann machines with hidden units. The only difference lies in the clamped phase of an iteration of the learning algorithm, where no longer all units are clamped by the environment. Now, an example only clamps the environmental units. After an example is clamped to the environmental units, the hidden units are allowed to reach equilibrium before the activation probabilities are estimated. Theorem 11.6 implies that the gradient of the divergence, as defined above, can still be calculated locally.

Theorem 11.6 *The partial derivative of the divergence $D(\mathbf{v}|\mathbf{v}')$, as given by (11.22), with respect to a connection strength $s_{\{u,v\}}$ is given by*

$$\frac{\partial D(\mathbf{v}|\mathbf{v}')}{\partial s_{\{u,v\}}} = -\frac{1}{c}\left[p'_{\{u,v\}}(c) - p_{\{u,v\}}(c)\right], \qquad (11.23)$$

where $p_{\{u,v\}}(c)$ denotes the probability that $\{u, v\}$ is activated when the Boltzmann machine has reached equilibrium and no units are clamped (free phase), and $p'_{\{u,v\}}(c)$ denotes the probability that $\{u, v\}$ is activated when the environmental units are clamped according to a desired visible behaviour $\mathbf{v}'(c)$ and the hidden units have reached equilibrium (clamped phase). ∎

To prove this theorem we need some additional definitions.

Definition 11.7 Let $\mathbf{X}$ and $\mathbf{Y}$ be stochastic variables whose values are, respectively, the configurations and environmental configurations of a Boltzmann machine, then the following probabilities can be defined:

$$v'_l(c) = \mathbb{P}_c^{(clamped)}\{\mathbf{Y} = l\}, \qquad (11.24)$$

$$v_l(c) = \mathbb{P}_c^{(free)}\{\mathbf{Y} = l\}, \qquad (11.25)$$

$$q'_k(c) = \mathbb{P}_c^{(clamped)}\{\mathbf{X} = k\}, \qquad (11.26)$$

$$q_k(c) = \mathbb{P}_c^{(free)}\{\mathbf{X} = k\}, \qquad (11.27)$$

$$p'_{\{u,v\}}(c) = \sum_{k\in\mathcal{R}} k(u)\,k(v)\,\mathbb{P}_c^{(clamped)}\{\mathbf{X} = k\},\text{ and} \qquad (11.28)$$

$$p_{\{u,v\}}(c) = \sum_{k\in\mathcal{R}} k(u)\,k(v)\,\mathbb{P}_c^{(free)}\{\mathbf{X} = k\}. \qquad (11.29)$$

∎

Furthermore, we need the following lemma.

Lemma 11.5 *From the probabilities of Definition 11.7 we can derive that*

$$q'_k(c) = \frac{v'_l(c)}{v_l(c)} \; q_k(c), \tag{11.30}$$

where $l = \mathcal{L}_\mathcal{V}(k)$ denotes the projection of k onto the environmental configuration space $\mathcal{V}$.

Proof To prove this lemma we recall that

$$
\begin{aligned}
q'_k(c) &= \mathbb{P}_c^{(clamped)}\{X = k\} \\
&= \mathbb{P}_c^{(clamped)}\{X = k | Y = l\} \; \mathbb{P}_c^{(clamped)}\{Y = l\} \\
&= \mathbb{P}_c^{(clamped)}\{X = k | Y = l\} \; v'_l(c). \tag{11.31}
\end{aligned}
$$

Similarly, we have

$$
\begin{aligned}
q_k(c) &= \mathbb{P}_c^{(free)}\{X = k\} \\
&= \mathbb{P}_c^{(free)}\{X = k | Y = l\} \; \mathbb{P}_c^{(free)}\{Y = l\} \\
&= \mathbb{P}_c^{(free)}\{X = k | Y = l\} \; v_l(c). \tag{11.32}
\end{aligned}
$$

Clearly, since in the clamped phase only the environmental units are clamped, we have

$$\mathbb{P}_c^{(clamped)}\{X = k | Y = l\} = \mathbb{P}_c^{(free)}\{X = k | Y = l\} \tag{11.33}$$

Combining (11.31), (11.32), and (11.33) now yields the proof of the lemma. ∎

We now prove Theorem 11.6 as follows. The partial derivative of $D(\mathbf{v}|\mathbf{v}')$ is given by

$$\frac{\partial D(\mathbf{v}|\mathbf{v}')}{\partial s_{\{u,v\}}} = -\sum_{l \in \mathcal{V}} \frac{v'_l(c)}{v_l(c)} \frac{\partial v_l(c)}{\partial s_{\{u,v\}}}.$$

The partial derivative of visible behaviour $v_l(c)$ with respect to $s_{\{u,v\}}$ is given by

$$\frac{\partial v_l(c)}{\partial s_{\{u,v\}}} = \sum_{k \in \mathcal{R}_{\mathcal{V}(l)}} \frac{\partial q_k(c)}{\partial s_{\{u,v\}}}.$$

From (11.9) we then obtain

$$\frac{\partial v_l(c)}{\partial s_{\{u,v\}}} = \frac{1}{c} \left[\sum_{k \in \mathcal{R}_{\mathcal{V}(l)}} k(u) \, k(v) \, q_k(c) - \sum_{k \in \mathcal{R}_{\mathcal{V}(l)}} p_{\{u,v\}}(c) \, q_k(c) \right]$$

$$= \frac{1}{c} \left[\sum_{k \in \mathcal{R}_{v(l)}} k(u) \, k(v) \, q_k(c) - p_{\{u,v\}}(c) \, v_l(c) \right].$$

(11.34)

Using (11.34) we obtain

$$\frac{\partial D(\mathbf{v}|\mathbf{v}')}{\partial s_{\{u,v\}}} = -\frac{1}{c} \sum_{l \in \mathcal{V}} \frac{v_l'(c)}{v_l(c)} \left[\sum_{k \in \mathcal{R}_{v(l)}} k(u) \, k(v) \, q_k(c) - p_{\{u,v\}}(c) \, v_l(c) \right].$$

Next, by using $\sum_{l \in \mathcal{V}} \sum_{k \in \mathcal{R}_{v(l)}} = \sum_{k \in \mathcal{R}}$ and $\sum_{l \in \mathcal{V}} v_l(c) = 1$, we derive that

$$\frac{\partial D(\mathbf{v}|\mathbf{v}')}{\partial s_{\{u,v\}}} = -\frac{1}{c} \left[\sum_{k \in \mathcal{R}} \frac{v_l'(c)}{v_l(c)} \, k(u) \, k(v) \, q_k(c) - p_{\{u,v\}}(c) \right].$$

Finally, from Lemma 11.4 we obtain

$$\frac{\partial D(\mathbf{v}|\mathbf{v}')}{\partial s_{\{u,v\}}} = -\frac{1}{c} \left[p'_{\{u,v\}}(c) - p_{\{u,v\}}(c) \right],$$

which completes the proof of Theorem 11.6. ∎

From Theorem 11.6 we conclude that for Boltzmann machines with hidden units the simple gradient method, discussed in Section 11.3, can also be used to minimize the divergence $D(\mathbf{v}|\mathbf{v}')$. However, the divergence $D(\mathbf{v}|\mathbf{v}')$ of the visible behaviour of a Boltzmann machine with respect to a desired visible behaviour cannot be proved to be convex. For Boltzmann machines with hidden units the Hessian matrix H which is given by the second partial derivatives of $D(\mathbf{v}|\mathbf{v}')$ is not positive semidefinite. A component of H can be shown to be equal to

$$\frac{\partial^2 D(\mathbf{v}|\mathbf{v}')}{\partial s_{\{u,v\}} \partial s_{\{w,z\}}} = \frac{1}{c^2} \left[\mathrm{Cov}_c(\mathbf{Z}_{\{u,v\}}, \mathbf{Z}_{\{w,z\}}) - \mathrm{Cov}_c(\mathbf{Z}'_{\{u,v\}}, \mathbf{Z}'_{\{w,z\}}) \right]$$

(11.35)

where $\mathbf{Z}_{\{u,v\}}$ and $\mathbf{Z}'_{\{u,v\}}$ are stochastic variables, defined by

$$\mathbf{Z}_{\{u,v\}} = \begin{cases} 1 & \text{if } \{u, v\} \text{ is activated in the free phase} \\ 0 & \text{if } \{u, v\} \text{ is not activated in the free phase} \end{cases}$$

(11.36)

$$\mathbf{Z}'_{\{u,v\}} = \begin{cases} 1 & \text{if } \{u, v\} \text{ is activated in the clamped phase} \\ 0 & \text{if } \{u, v\} \text{ is not activated in the clamped phase.} \end{cases}$$

(11.37)

From this definition it can be shown that, in general, there exist $s \in \mathbb{R}^{|C|}$ for which $s^T H s < 0$. Hence, the divergence $D(v|v')$ generally has local minima for which $D(v|v') \neq 0$. However, this does not mean that the learning algorithm necessarily gets trapped in the first local minimum which is encountered. Due to the fact that the activation probabilities are approximated the learning algorithm does not converge strictly, i.e. the method is not stable. In some iterations the connection strengths are adjusted such that the divergence increases. On average, the generated sequence $\{s^{(k)}\}$ moves in the direction of a decreasing divergence, but due to its stochastic nature, steps that increase the divergence may frequently occur. This implies that the learning algorithm can escape from local minima in very much the same way as the simulated annealing algorithm escapes from local minima.

It should be clear that increasing the number of samples from which the relative activation frequencies $z'_{\{u,v\}}$ and $z_{\{u,v\}}$ are calculated, results in a better approximation of the direction of the steepest descent. However, it increases the probability of getting stuck in poor local minima of the divergence. In practice, a trade-off has to be made between the speed of convergence and the probability of ending up in a poor local minimum. We end this section with an example.

Example 11.3 (Exclusive-or problem) Let us revisit the example of the exclusive-or problem. In Example 10.2 we have seen that the exclusive-or function cannot be implemented on a Boltzmann machine without using hidden units. Similarly to Example 10.2, we again introduce a hidden unit. Now, the Boltzmann machine can learn the exclusive-or function, such that the four environmental configurations shown in Figure 10.3a each correspond to a global maximum of the consensus function C. These four global maxima should of course be the only global maxima. Note that in this example we are not interested in learning a specific visible behaviour, but satisfy ourselves with the requirement that the four global maxima are unique.

The four globally maximal configurations can be chosen in a number of different ways, by assigning to each of the environmental configurations shown in Table 10.3a, an internal configuration, which is determined by the state of the single hidden unit. This gives 16 possibilities which are given in Table 11.2. Of these 16 possible internal representations, 6 can be shown to lead to conflicting requirements on the connection strengths. As an example we discuss one of these infeasible internal representations, viz. the one that corresponds to the third column of the infeasible internal representations in Table 11.2. The requirement that the four configurations $a = (0000)$,

input $(\mathcal{I}')$		output $(\mathcal{O}')$	hidden $(\mathcal{H}')$	
			feasible	*infeasible*
0	0	0	0 0 0 0 0 1 1 1 1 1	0 0 0 1 1 1
0	1	1	0 0 1 1 1 0 0 0 1 1	0 0 1 0 1 1
1	0	1	0 1 0 1 1 0 0 1 0 1	0 1 0 1 0 1
1	1	0	1 0 0 0 1 0 1 1 1 0	0 1 1 0 0 1

Table 11.2: The possible internal representations for a Boltzmann machine with one hidden unit.

$b = (0111)$, $c = (1010)$, and $d = (1101)$ must be the only global maxima of the consensus function C leads to the following contradiction: if a is a maximum, then $\Delta C_a(u_1) < 0$, so that

$$s_{\{u_1,u_1\}} < 0. \tag{11.38}$$

If b is a maximum, then $\Delta C_b(u_1) < 0$, hence

$$s_{\{u_1,u_1\}} + s_{\{u_1,u_2\}} + s_{\{u_1,u_3\}} + s_{\{u_1,u_4\}} < 0. \tag{11.39}$$

If c is a maximum, then $\Delta C_c(u_1) < 0$, which gives

$$s_{\{u_1,u_1\}} + s_{\{u_1,u_3\}} > 0. \tag{11.40}$$

Combining (11.39) and (11.40) yields

$$s_{\{u_1,u_2\}} + s_{\{u_1,u_4\}} < 0. \tag{11.41}$$

Combining (11.38) and (11.41) we obtain

$$s_{\{u_1,u_1\}} + s_{\{u_1,u_2\}} + s_{\{u_1,u_4\}} < 0. \tag{11.42}$$

Furthermore, if d is maximum, then $\Delta C_d(u_1) < 0$, which gives

$$s_{\{u_1,u_1\}} + s_{\{u_1,u_2\}} + s_{\{u_1,u_4\}} > 0. \tag{11.43}$$

Clearly, (11.43) contradicts (11.42).

Sejnowski, Kienker & Hinton [1986] have run the learning algorithm on the exclusive-or problem 339 times. The learning algorithm obtained several different internal representations. From the ten feasible internal representations eight were found by these simulations. Table 11.3 gives the relative

input	output	feasible internal representations							
0 0	0	0	0	0	0	1	1	1	1
0 1	1	0	0	1	1	0	0	1	1
1 0	1	0	1	0	1	0	1	0	1
1 1	0	1	0	0	1	0	1	1	0
rel. frequencies:		7%	16%	16%	4%	54%	1%	1%	0.3%

Table 11.3: The internal representations found by running the learning algorithm 339 times, together with their relative frequencies [Sejnowski, Kienker & Hinton, 1986].

frequencies of occurring of the internal representations. The fact that some internal representations are found more frequently than others can be explained from the choice of the initial values of the connection strengths [Sejnowski, Kienker & Hinton, 1986]. ∎

11.5   Variants of the Learning Algorithm

Before discussing some practical aspects of learning on a Boltzmann machine, we give some variants of the learning algorithm.

In Table 11.4 a number of related learning approaches are given that are basically only minor variants of the learning algorithm given in Section 11.2. For each approach it is indicated whether units are clamped or equilibrated in the clamped and the free phase. The general learning algorithm, as discussed in Sections 11.3 and 11.4, is given in the first and second row, for Boltzmann machines with or without hidden units, respectively.

If we are only interested in strict classification, i.e. in recognizing given inputs only, the input units can always be clamped. Usually, this considerably accelerates the learning process. If more units run free, then it takes more time to reach equilibrium. Thus, if the input units are clamped in the free phase - according to the desired global behaviour - the free phase requires less time. This approach is especially of interest when classification problems require many input units, which is generally the case for image and speech recognition [Prager, Harrison & Fallside, 1986]. This strict classification approach is given in rows three and four of Table 11.4, for Boltzmann machines with and without hidden units, respectively.

	clamped phase	free phase
general formulation with hidden units (Section 11.3)	environmental units are clamped hidden units equilibrate	all units equilibrate
general formulation without hidden units (Section 11.2)	all units are clamped	all units equilibrate
strict classification with hidden units	environmental units are clamped hidden units equilibrate	input units are clamped other units equilibrate
strict classification without hidden units	all units are clamped	input units are clamped output units equilibrate
associative or content addressable memories	input units are clamped hidden units equilibrate	all units equilibrate

Table 11.4: For different variants of the learning algorithm it is indicated how the units behave in the clamped and free phases of each iteration.

Up to now, we have only discussed learning applications within the context of classification. A closely related application area, for which the learning capabilities of Boltzmann machines may be of interest, is the area of *associative or content-addressable memories* [Kohonen, 1987; 1988]. Associative memories are used to store a number of patterns. After learning these patterns, the associative memory is used to complete a partially given pattern. This problem can easily be formulated in terms of the Boltzmann machine model. Here, a Boltzmann machine is thought to consist of input and hidden units only. Patterns are learned by clamping them as examples to the input units; see row five of Table 11.4. When learning, the connection strengths are adjusted such that the patterns correspond to local maxima of the consensus function. By clamping some of the input units, the most probable completion of this partial pattern is generated by maximizing the consensus. For this application, no classification labels are given along with the presented patterns. Therefore, this learning approach is often referred to as *unsupervised learning*, while the approach for classification purposes is called *supervised learning*. Note that for the general formulation of the learning algorithm there are no essential differences between supervised and unsupervised learning.

11.6   Learning in Practice

In this section we discuss some practical implementation aspects of the learning algorithm. The implementation aspects are illustrated by Example 11.4, given at the end of the section.

11.6.1   Choosing a Desired Visible Behaviour

Firstly, we have to specify a visible behaviour $\mathbf{v}'(c)$, that gives the desired probability of occurring at equilibrium for each environmental configuration l. Clearly, we want $v_l'(c)$ to be large for each environmental configuration l in the learning set $\mathcal{V}'$. In order to be feasible $\mathbf{v}'(c)$ should at least be chosen such that $0 < v_l'(c) < 1$ for all $l \in \mathcal{V}$. To choose the desired visible behaviour, we make the following observation. The connection strengths required to implement a visible behaviour $\mathbf{v}'(c)$ are generally larger whenever $\mathbf{v}'(c)$ is less smooth. A desired visible behaviour $\mathbf{v}'(c)$ is called *smooth* if for every pair of environmental configurations l and m the desired probabilities of occurring, $v_l'(c)$ and $v_m'(c)$, are approximately equal if l and m are approximately equal. The above observation is explained as follows. If l and m differ only in one unit, the difference in consensus is caused by (de)activation of only a few connections. If the corresponding probabilities $v_l'(c)$ and $v_m'(c)$ are substantially different the strengths of the connections that are (de)activated must be large. Thus, generally speaking, the smoother the desired visible behaviour, the smaller the strengths necessary to implement this behaviour. In the extreme case, if all environmental configurations have equal probabilities of occurring, then all connection strengths can be chosen equal to zero.

To avoid large connection strengths we should choose a reasonably smooth desired behaviour $\mathbf{v}'(c)$. However, choosing $\mathbf{v}'(c)$ too smooth makes a correct classification, after learning, more difficult. Clearly, there is a trade-off between the ease of learning and the ease of classification afterwards.

In practice, one obtains a reasonably smooth visible behaviour $\mathbf{v}'(c)$ as follows. Only the environmental configurations in the learning set $\mathcal{V}'$ are clamped to the environmental units (with equal probabilities). If such a configuration is clamped to the units, the state of each environmental unit is changed with a probability p, before the hidden units equilibrate. Note that, in this way, all environmental configurations have a positive probability to occur as examples. The resulting visible behaviour is smoother, whenever probability p is chosen larger.

11.6.2 Convergence Properties

In Section 11.2 we mentioned that the speed of convergence of the iterative method, used to minimize the divergence $D(\mathbf{q}|\mathbf{q}')$, is smaller than that of the steepest descent method that moves in the same direction but uses larger steps. In the literature on nonlinear optimization [Abadie, 1970; Luenberger, 1973] the steepest descent method is commonly referred to as being, though theoretically interesting, of little practical value, due to its poor convergence properties. Clearly, as the speed of convergence of the learning algorithm is even worse, further research should aim at improving it. Many alternatives to the steepest descent method that yield better convergence properties have been proposed in the literature. We do not consider using these methods, however, as they violate the locality principle.

In practice, one often uses an iterative method in which the strengths of the connections are adjusted according to

$$s_{\{u,v\}}^{(i+1)} = s_{\{u,v\}}^{(i)} - f\left(\frac{\partial D(\mathbf{v}|\mathbf{v}')}{\partial s_{\{u,v\}}}\right), \tag{11.44}$$

where $f(x)$ is defined as

$$f(x) = \begin{cases} \beta & x > 0 \\ 0 & x = 0 \\ -\beta & x < 0, \end{cases} \tag{11.45}$$

for some constant $\beta \in \mathbb{R}^+$. Thus, connection strengths are only changed by a fixed amount β. Note that here the algorithm no longer moves in the direction of the steepest descent. Little is known about the convergence properties of this method. However, several authors have reported that they achieve good results with this approach in less time than the gradient method [Aarts & Korst, 1986; Ackley, Hinton & Sejnowski, 1985; Hinton, Sejnowski & Ackley, 1984; Prager, Harrison & Fallside, 1986]. An explanation for this might be the following. The slope of the divergence $D(\mathbf{v}|\mathbf{v}')$ is often very small, which results in a very slow convergence of the gradient method. For these situations, the alternative method yields a faster convergence.

11.6.3   Estimation of the Activation Probabilities

For Boltzmann machines with hidden units, the divergence $D(\mathbf{v}|\mathbf{v}')$ usually has one or more local minima. In Section 11.3 we have already explained the advantages and disadvantages of an accurate estimation of the activation probabilities. The more accurate the estimation, the larger the speed of convergence of the sequence $\{\mathbf{s}^{(k)}\}$. However, an accurate estimation of the gradient requires large computation efforts for equilibration. Furthermore, it increases the probability of getting stuck in a poor local minimum. The accuracy of the estimations is determined by

-   the number of examples that are clamped during each iteration of the learning algorithm,

-   the extend to which the Boltzmann machine is equilibrated, and

-   the sample size from which the activation probabilities are estimated after equilibration.

In the literature these choices are often only empirically motivated [Bounds, 1986a; 1986b; Hinton, Sejnowski & Ackley, 1984; Prager, Harrison & Fallside, 1986].

Clearly, equilibration in a Boltzmann machine can only be reached asymptotically; see Chapter 8. Therefore, we must use a finite-time implementation. Equilibration of the free units can be carried out at a fixed value of the control parameter c. However, it is more efficient to equilibrate at descending values of c. Then equilibration can again be implemented as a sequence of Markov chains of finite length. The initial value of the control parameter $c_0^{(u)}$ for a unit u is chosen such that the acceptance ratio is close to 0.5. At the end of each Markov chain the value of the control parameter is lowered by $c_{j+1}^{(u)} = \alpha c_j^{(u)}$, for some $\alpha \in \mathbb{R}^+$. Equilibration is stopped if the acceptance ratio is approximately 0.25. The length L of each Markov chain is chosen proportional to the number of unclamped units.

11.6.4   Termination of the Learning Algorithm

The learning process can be terminated using only locally available information as follows. A connection $\{u, v\}$ stops estimating the activation probabilities if $|z_{\{u,v\}} - z'_{\{u,v\}}| < \epsilon$ for K subsequent iterations of the learning algorithm.

To illustrate the practical learning aspects discussed in this section we readdress the seven-segment display problem; see Section 10.3.1.

Example 11.4 (Seven-segment display) For this problem a Boltzmann machine is chosen with 7 input units, 10 hidden units and 10 output units. Now, we want the Boltzmann machine to learn to recognize correctly the digits given in Figure 10.1. Note that the number of hidden units guarantees that the required input-output behaviour can be implemented; see Theorem 10.2. The units are interconnected as specified in Section 10.4.3 and all connection strengths are initially chosen equal to 0.

The set of environmental configurations $\mathcal{V}'$ that the Boltzmann machine must learn, i.e. the learning set, is given in Table 10.1. The learning set plays a similar role as the classification set in Chapter 10. An example is presented to the Boltzmann machine by randomly choosing an environmental configuration from $\mathcal{V}'$ and clamping the environmental units accordingly. Then the state of each environmental unit is changed with a probability $p = 0.05$, before equilibration is started. This guarantees a reasonably smooth desired visible behaviour $\mathbf{v}'$.

The iterative approach discussed in Section 11.6.2 is used where connection strengths are adjusted by a fixed amount of β. Here, β is chosen equal to 1. For our purposes the number of examples that during each iteration is clamped to the environmental units, is chosen equal to 20. Equilibration is carried out as discussed in Section 11.6.3, where α is chosen equal to 0.8 and the length of a Markov chain is chosen as twice the number of unclamped units. A connection $\{u, v\}$ stops estimating the activation probabilities if $|z_{\{u,v\}} - z'_{\{u,v\}}| < 0.05$ for 10 subsequent iterations of the learning algorithm. For the seven-segment display problem, the learning algorithm stops after 870 iterations of the learning algorithm. The Boltzmann machine is then tested as follows.

Firstly, the Boltzmann machine is tested by clamping only the input units to one of the 'known' inputs, i.e. the inputs of the learning set, and letting the Boltzmann machine maximize its consensus by adjusting the states of the hidden and output units. Each input is presented to the Boltzmann machine 100 times. The results of this test are shown in Table 11.5. We observe that a correct output is generated for 972 of the 1000 presented inputs. In the case of an incorrect classification, the Boltzmann machine tends to generate an output that corresponds to a digit that has a large similarity to the presented digit, e.g. 1 instead of 7, and 0 instead of 8. A large similarity can here be defined as having a small Hamming distance; see Table 11.6.

Secondly, the Boltzmann machine is tested by clamping only the output units to one of the 'known' outputs. Each output is presented 100 times. In this case, the machine always finds the correct input.

	0	1	2	3	4	5	6	7	8	9
0	100	-	-	-	-	-	-	-	-	-
1	-	98	-	-	-	-	-	2	-	-
2	-	-	99	1	-	-	-	-	-	-
3	-	-	-	100	-	-	-	-	-	-
4	-	-	-	-	100	-	-	-	-	-
5	-	-	-	-	-	100	-	-	-	-
6	-	-	-	-	-	-	100	-	-	-
7	-	13	-	-	-	-	-	87	-	-
8	7	-	2	-	-	-	-	-	91	-
9	-	-	-	-	3	-	-	-	-	97

Table 11.5: Score matrix for the test with the 10 inputs of the learning set.

To test the associative capabilities of the Boltzmann machine we also presented the 'unknown' inputs, given in Table 10.4. Again these inputs are presented to the Boltzmann machine 100 times each. The results of this test are shown in Table 11.7. The corresponding Hamming distances are given in Table 11.8. For 900 of the 1000 presented inputs an output is generated whose corresponding input has a Hamming distance 1 to the presented input. ■

11.7   Robustness Aspects

Robustness is generally considered an important property of connectionist models. The correct performance of a conventional computer strongly depends on the correct performance of all its components, requiring stringent demands on their reliability. Due to the use of distributed representations, a correct performance of a connectionist network model is less dependent on the correct performance of all its components. Failure of a small number of components may still result in correct overall performance.

We briefly mention two interesting phenomena illustrating the robustness of the connection strengths generated by the learning algorithm.

	0	1	2	3	4	5	6	7	8	9
0	0	4	3	3	4	3	3	3	1	3
1	4	0	5	3	2	5	5	1	5	3
2	3	5	0	2	5	4	4	4	2	4
3	3	3	2	0	3	2	4	2	2	2
4	4	2	5	3	0	3	3	3	3	1
5	3	5	4	2	3	0	2	4	2	2
6	3	5	4	4	3	2	0	6	2	4
7	3	1	4	2	3	4	6	0	4	2
8	1	5	2	2	3	2	2	4	0	2
9	3	3	4	2	1	2	4	2	2	0

Table 11.6: The Hamming distances between the different inputs. The Hamming distance between two inputs equals the number of segments or units in which they differ.

11.7.1 Internal Representations

When using hidden units, a Boltzmann machine constructs an internal representation which captures the regularities in the desired visible behaviour. To that end, a correspondence between environmental and internal configurations is constructed, as we have seen in Example 11.3.

Hinton, Sejnowski & Ackley [1984] describe a number of experiments that specifically focus on the construction of such internal representations. From these experiments they reach the following conclusions. If the number of possible internal configurations is relatively large as compared to the cardinality of the learning set, then the Boltzmann machine tends to construct an internal representation, such that the internal configurations that are used have large mutual Hamming distances. This phenomenon makes the machine insensitive to minor physical damage. If the internal configurations that are used all have large mutual Hamming distances, then a malfunctioning of one or more hidden units cannot seriously affect the Boltzmann machine's performance.

For further details the reader is referred to Ackley, Hinton & Sejnowski [1985].

	0	1	2	3	4	5	6	7	8	9
⌐	82	-	-	-	-	-	18	-	-	-
ı	-	100	-	-	-	-	-	-	-	-
?	-	-	100	-	-	-	-	-	-	-
̅ɔ	-	-	2	56	-	42	-	-	-	-
⅄	-	-	-	-	97	-	-	-	-	3
⌐ı	30	-	-	7	-	61	2	-	-	3
ப	7	-	-	-	-	-	93	-	-	-
̅ı	-	1	-	-	-	-	-	99	-	-
∂	4	-	49	35	-	-	-	-	12	-
⅂	-	26	-	-	3	-	-	64	-	10

Table 11.7: Score matrix for the test with the 10 'unknown' inputs.

11.7.2   Relearning

If a Boltzmann machine has learned some global visible behaviour, we can examine the effect of adding noise to the connection strengths. Changing the strengths by adding uniform random noise in a certain range $(-x\%, +x\%)$ causes a significant deterioration of the performance of the Boltzmann machine, if x is chosen to be large. However, the performance can be improved again to almost the original performance by a short process of relearning. By relearning we mean a restart of the learning algorithm with the distorted set as the initial set of connection strengths. The time needed to almost recover the original performance is usually only a small fraction of the time needed for learning [Aarts & Korst, 1988b; Hinton & Sejnowski, 1986]. This phenomenon can be understood as follows. By adding random noise, the global shape of the consensus function is not seriously affected. Little effort is therefore needed to recover the original shape. Furthermore, due to the distributed character of the internal representation, all examples used during the process of relearning tend to change the connection strengths in the same direction, to the original set of strengths. For further details on this subject the reader is referred to Hinton & Sejnowski [1986].

	0	1	2	3	4	5	6	7	8	9
0	1	5	4	4	5	2	2	4	2	4
1	5	1	4	4	3	6	6	2	6	4
2	4	4	1	3	4	5	5	3	3	3
3	4	4	3	1	4	1	3	3	3	3
4	5	3	6	4	1	2	2	4	4	2
5	2	4	5	3	4	1	3	3	3	3
6	2	4	5	5	4	3	1	5	3	5
7	4	2	5	3	4	3	5	1	5	3
8	2	4	1	1	4	3	3	3	1	3
9	2	2	5	3	2	3	5	1	3	1

Table 11.8: The Hamming distances between the 'known' and the 'unknown' inputs.

11.8 Discussion

Several simulations of small-scale learning problems have been discussed in the literature [Aarts & Korst, 1986; Ackley, Hinton & Sejnowski, 1985; Hinton & Sejnowski, 1986; Sejnowski, Kienker & Hinton, 1986]. The most successful application of the Boltzmann machine to practical learning problems has been in the field of speech-pattern recognition [Bridle & Moore, 1984; Prager, Harrison & Fallside, 1986; Trehern, Jack & Laver, 1986]. Prager, Harrison & Fallside present large-scale computer simulations that demonstrate the ability of a 2000-unit Boltzmann machine to distinguish between the 11 steady-state vowels in English with an accuracy of 85%. Both static and time-varying speech patterns could be recognized, and from their extensive set of numerical results they arrived at the conclusion that Boltzmann machines are suitable for connected speech recognition.

The average-case performance and time complexity of the learning algorithm are very important, but they are also very complex owing to the many parameters involved. We mention

1. the parameters of the learning algorithm (cooling schedule, adjustment of the strengths, smoothness of the desired visible behaviour, number of examples per iteration),

2. the ratio between the environmental and the hidden units,

3. the architectural structure (connection pattern), and

4. the intrinsic difficulty of a given problem in terms of potentially conflicting requirements on the connection strengths.

Most of the investigations presented in the literature concentrate on the first item [Aarts & Korst, 1986; Bounds, 1986a; Derthick, 1984; Prager, Harrison & Fallside, 1986]. Not much is known in the literature about the items (2)-(4). Hinton, Sejnowski & Ackley [1984] discuss some properties on the basis of the examples they investigated, but they did not arrive at more general conclusions. Prager, Harrison & Fallside [1986] investigated Boltzmann machine configurations tailored to speech processing. For these applications they conclude that a Boltzmann machine should have a large number of input units into which data can be clamped using a neighbourhood code.

The overall conclusion from the various numerical simulation experiments reported in the literature is that learning in a Boltzmann machine is slow. Two reasons can be identified for the slow convergence of the learning algorithm.

- The first and probably most essential reason is the time spent for equilibration. This problem is inherent to the stochastic nature of the Boltzmann machine. Compared to the backpropagation networks introduced by Rumelhart, Hinton & Williams [1986], learning in a Boltzmann machine is more than one order of magnitude slower, while producing similar results for many applications [Hinton, 1987]. However, the model of the Boltzmann machine is more general and can be used for some applications that cannot be dealt with by backpropagation networks. Furthermore, backpropagation networks seem to be less well suited to be put directly onto silicon. Alspector & Allen [1987] give a number of reasons in support of this. One of the most important is that, during learning in a backpropagation network global information is used, which implies an inefficient use of the potential parallelism of a hardware implementation.

- The convergence properties of the gradient method used in the learning algorithm are poor; see Section 11.6.2.

There are several approaches that might lead to a speed-up of the learning algorithm.

- Acceleration through hardware implementation of the Boltzmann machine. Most experiments with Boltzmann machines presented in the literature are based on sequential simulations on conventional computer systems. This is probably the reason why only relatively small-scale examples have been considered. A number of authors indicate a possible hardware implementation of the Boltzmann machine, either by a VLSI silicon implementation [Alspector & Allen, 1987] or by an optoelectronic implementation [Farhat, 1987; Ticknor & Barrett, 1987]; see also Section 9.8.

- Higher-order Boltzmann machines. Sejnowski [1986] argues that Boltzmann machines, in which multiples of units interact through symmetric conjunctive connections, such that the consensus function can be expressed as a higher-order polynomial of the states of the units, should yield a much faster learning rate than the conventional Boltzmann machines based on pairwise interactions.

- Improvement of the iterative method used to minimize the divergence $D(\mathbf{v}|\mathbf{v}')$. The second derivative of the divergence can be used to select directions that yield faster convergence than the direction of the steepest descent. This improvement, however, complicates the calculations since the second derivatives can no longer be calculated locally [Derthick, 1984], which is cumbersome for hardware implementations.

- Decomposition of the Boltzmann machine into a modular hierarchically structured network in which the modules are weakly coupled. Clearly, this is the appropriate way to tackle complexity. However, finding appropriate decompositions is a very difficult task. Some recent discussions on this subject are presented by Ballard [1987] and Hinton [1987].

The poor convergence properties of learning in Boltzmann machines has, up to the present, prevented application of the learning capabilities of Boltzmann machines to large practical problems. Hardware implementations that effectively exploit the inherent parallelism of the model will, of course, significantly speed-up learning.

The use of hardware implementations, which will probably clear the way for practical use of Boltzmann machines within application areas such as combinatorial optimization and classification, is probably not the final answer for learning applications. Further research on alternative learning algorithms for the Boltzmann machine is needed to guarantee its practical use.

APPENDIX

The EUR100 Problem Instance

The EUR100 problem instance [Aarts, Korst & Van Laarhoven, 1988] is a symmetric, Euclidean instance of the travelling salesman problem, defined on 100 major European cities (see Table A.1).

1	Amsterdam	26	Dortmund	51	London	76	Roma
2	Antwerpen	27	Dresden	52	Luxembourg	77	Rostock
3	Athínai	28	Dublin	53	Lyon	78	Rotterdam
4	Barcelona	29	Düsseldorf	54	Madrid	79	Sarajevo
5	Basel	30	Edinburgh	55	Magdeburg	80	Sevilla
6	Belfast	31	Gdansk	56	Malaga	81	Sheffield
7	Beograd	32	Genova	57	Malmö	82	Skopje
8	Bergen	33	Glasgow	58	Manchester	83	Smolensk
9	Berlin	34	Göteborg	59	Marseille	84	Sofija
10	Bern	35	Granada	60	Milano	85	Southampton
11	Bilbao	36	Graz	61	Minsk	86	Split
12	Birmingham	37	Hamburg	62	Monaco	87	Stockholm
13	Bonn	38	Hannover	63	Moskva	88	Strasbourg
14	Bordeaux	39	Helsinki	64	München	89	Stuttgart
15	Bratislava	40	Istanbul	65	Napoli	90	Thessaloniki
16	Bremen	41	Köln	66	Nice	91	Torino
17	Brno	42	København	67	Odessa	92	Toulouse
18	Bruxelles	43	Krakow	68	Oslo	93	Trieste
19	Bucureşti	44	Leeds	69	Palermo	94	Turku
20	Budapest	45	Leipzig	70	Paris	95	Uppsala
21	Burgas	46	Leningrad	71	Plovdiv	96	Valencia
22	Constanta	47	Liège	72	Plzen	97	Warszawa
23	Cork	48	Lisboa	73	Pôrto	98	Wien
24	Craiova	49	Liverpool	74	Praha	99	Zagreb
25	Den Haag	50	Lódz	75	Riga	100	Zürich

Table A.1: The 100 European cities in the EUR100 problem instance.

The elements of the distance matrix are calculated from the geographic co-ordinates of the cities given in the table. The coordinates are taken from the *Times Atlas*. The distance between two cities is taken as the length of the cord between two points (cities) on an ellipsoid with axes a=6378 and b=6357. The polar angles of the points are given by the coordinates mentioned above. A listing of the distance matrix is available from the authors.

Figure A.1 shows the location of the 100 cities, together with a minimal tour, 21134 km in length.

The minimum tour length is calculated using a tailored optimization algorithm developed at the University of Amsterdam by Jonker & Volgenant [1984]. The algorithm uses a branch and bound technique based on a 1-tree relaxation, applying a mechanism that eliminates non-optimal edges of the problem instance. A minimal tour was obtained in 59.5 CPU seconds on a CYBER-205 computer.

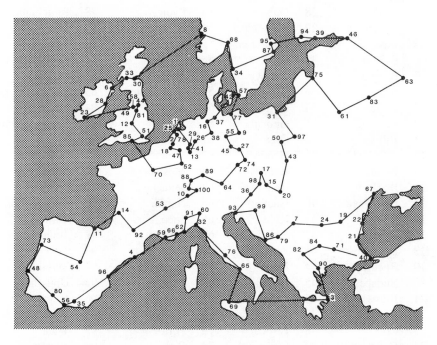

Figure A.1: The EUR100 travelling salesman problem. The solid lines indicate an optimal tour.

Bibliography

AARTS, E.H.L., G.F.M. BEENKER, AND J.H.M. KORST [1985], Asymptotic probability calculations for aselect sampling, *Philips Research Report 6073*.

AARTS, E.H.L., F.M.J. DE BONT, J.H.A. HABERS, AND P.J.M. VAN LAARHOVEN [1986], Parallel implementations of the statistical cooling algorithm, *Integration* **4**, 209-238.

AARTS, E.H.L., AND J.H.M. KORST [1986], Simulation of learning in parallel networks based on the Boltzmann machine, *Proc. 2nd European Simulation Congres*, Antwerp, 391-398.

AARTS, E.H.L., AND J.H.M. KORST [1987], Boltzmann machines and their applications, *Lecture Notes in Computer Science* **258**, Springer-Verlag, Berlin, 34-50.

AARTS, E.H.L., AND J.H.M. KORST [1988a], Boltzmann machines for travelling salesman problems, *European Journ. of Operational Research*, in press.

AARTS, E.H.L., AND J.H.M. KORST [1988b], Computations in massively parallel networks based on the Boltzmann machine: a review, *Parallel Computing*, in press.

AARTS, E.H.L., J.H.M. KORST, AND P.J.M. VAN LAARHOVEN [1988], A quantitative analysis of the simulated annealing algorithm: a case study for the traveling salesman problem, *Journ. of Statistical Physics* **50**, 189-206.

AARTS, E.H.L., AND P.J.M. VAN LAARHOVEN [1985a], Statistical cooling: a general approach to combinatorial optimization problems, *Philips Journ. of Research* **40**, 193-226.

AARTS, E.H.L, AND P.J.M. VAN LAARHOVEN [1985b], A new polynomial time cooling schedule, *Proc. IEEE Int. Conf. on Computer-Aided Design*, Santa Clara, 206-208.

AARTS, E.H.L., AND P.J.M. VAN LAARHOVEN [1987], Simulated annealing: a pedestrian review of the theory and some applications, in: P.A. Devijver and J. Kittler (Eds.), *Pattern Recognition Theory and Applications*, NASI Series on Computer and Systems Sciences **30**, Springer-Verlag, Berlin, 179-192.

AARTS, E.H.L., AND P.J.M. VAN LAARHOVEN [1988], Simulated annealing: An introduction, *Statistica Neerlandica* **43**.

ABADIE, J. (ED.) [1970], *Integer and Nonlinear Programming*, North-Holland, Amsterdam.

ABRAMSON, D. [1987], Constructing school timetables using simulated annealing: sequential and parallel algorithms, Royal Melbourne Institute of Technology, *Technical Report 112069R*.

ABU-MOSTAFA, Y.S. AND D. PSLATIS [1987], Optical neural computers, *Scientific American* **256**, 88-95.

ACKLEY, D.H., G.E. HINTON AND T.J. SEJNOWSKI [1985], A learning algorithm for Boltzmann machines, *Cognitive Science* **9**, 147-169.

AHO, A.V., J.E. HOPCROFT AND J.D. ULLMAN [1974], *The Design and Analysis of Computer Algorithms*, Addison-Wesley, Reading, (MA).

ALSPECTOR, J., AND R.B. ALLEN [1987], A neuromorphic VLSI learning system, in: P. Losleben (Ed.), *Advanced Research in VLSI*, MIT Press, Cambridge (MA), 313-349.

AMARI, S., AND M.A. ARBIB (EDS.) [1982], *Competition and Cooperation in Neural Networks*, Springer-Verlag, New York.

AMIT, D.J., H. GUTFREUND, AND H. SOMPOLINSKY [1987], Information storage in neural networks with low levels of activity, *Physical Review A* **35**, 2293-2303.

ANILY, S., AND A. FEDERGRUEN [1987a], Ergodicity in parametric nonstationary Markov chains: an application to simulated annealing methods, *Operations Research* **35**, 867-874.

ANILY, S., AND A. FEDERGRUEN [1987b], Simulated annealing methods with general acceptance probabilities, *Journ. of Applied Probability* **24**, 657-667.

BAGHERZADEH, N., T. KEROLA, B. LEDDY, AND R. BRICE [1987], On parallel execution of the travelling salesman problem on a neural network model, *Proc. Int. Conf. on Neural Networks*, San Diego, **3**, 317-324.

BAHL, R.L., F. JELINEK, AND R.L. MERCER [1983], A maximum likelihood approach to continuous speech recognition, *IEEE Trans. on Pattern Analysis and Machine Intelligence* **5**, 179-190.

BALLARD, D.H. [1987], Modular learning in neural networks, *Proc. National Conf. on AI, AAAI-87*, Seattle.

BALLARD, D.H., G.E. HINTON, AND T.J. SEJNOWSKI [1983], Parallel visual computation, *Nature* **306**, 21-26.

BANERJEE, P., AND M. JONES [1986], A parallel simulated annealing algorithm for standard cell placement on a hypercube computer, *Proc. IEEE Int. Conf. on Computer-Aided Design*, Santa Clara, 34-37.

BARKER, J.A., AND D. HENDERSON [1976], What is "liquid"? Understanding the states of matter, *Reviews of Modern Physics* **48**, 587-671.

BARR, A., AND E.A. FEIGENBAUM [1981], *The Handbook of Artificial Intelligence* **1**, Pitman Books, London.

BARR, A., AND E.A. FEIGENBAUM [1982], *The Handbook of Artificial Intelligence* **2**, Pitman Books, London.

BEENKER, G.F.M., T.A.C.M. CLAASEN, AND P.W.C. HERMENS [1985], Binary sequences with maximally flat amplitude spectrum, *Philips Journ. of Research* **40**, 289-304.

BERNASCONI, J., [1987], Low autocorrelation binary sequences: statistical mechanics and configuration space analysis, *Journ. de Physique* **48**, 559-567.

BHASKER, J., AND S. SAHNI [1987], Optimal linear arrangement of circuit components, *Journ. of VLSI and Computer Systems* **2**, 87-109.

BINDER, K. [1978], *Monte Carlo Methods in Statistical Physics*, Springer-Verlag, New York.

BLISS, T.V.P., AND T. LOMO [1973], Long lasting potentiation of synaptic transmission in the dentate area of the anesthetized rabbit following simulation of the perforant path, *Journ. of Physiology* **232**, 331-358.

BOHACHEVSKY, I., M.E. JOHNSON, AND M.L. STEIN [1986], Generalized simulated annealing for function optimization, *Technometrics* **28**, 209-217.

BONOMI, E., AND J.-L. LUTTON [1984], The N-city travelling salesman problem: statistical mechanics and the Metropolis algorithm, *SIAM Review* **26**, 551-568.

BONOMI, E., AND J.-L. LUTTON [1986], The asymptotic behaviour of quadratic sum assignment problems: a statistical mechanics approach, *European Journ. of Operational Research* **26**, 295-300.

BONT, F.M.J. DE, E.H.L. AARTS, P. MEEHAN, AND C.G. O'BRIEN [1988], Placement of shapeable blocks, *Philips Journ. of Research* **43**, 1-19.

BOSCH, L. TEN [1988], Learnable sets by Boltzmann machines, Nijmegen University KUN, Nijmegen, preprint.

BOUNDS, D.G. [1986a], Numerical simulations of Boltzmann machines, in: J.S. Denker (Ed.), *Neural Networks for Computing*, AIP Conf. Proc. **151**, Snowbird (UT), 59-64.

BOUNDS, D.G. [1986b], A statistical mechanical study of Boltzmann machines, *Journ. of Physics A* **20**, 2133-2145.

BRAITENBERG, V. [1978], Cell assemblies in the cerebral cortex, in: R. Heim and G. Palm (Eds.), *Lecture Notes in Biomathematics* **21**, Springer-Verlag, Berlin, 171-188.

BRIDLE, J.S., AND R.K. MOORE [1984], Boltzmann machines for speech pattern recognition, *Proc. IOA* **6**, 315-322.

BRUCK, J., AND J.W. GOODMAN [1987], A generalized convergence theorem for neural networks and its application to combinatorial optimization, *Proc. Int. Conf. on Neural Networks* **3**, San Diego, 649-656.

BURKARD, R.E., AND F. RENDL [1984], A thermodynamically motivated simulation procedure for combinatorial optimization problems, *European Journ. of Operational Research* **17**, 169-174.

CARBONELL, J.G., R.S. MICHALSKI, AND T.M. MITCHELL [1983], An overview of machine learning, in: R.S. Michalski, J.G. Carbonell, and T.M. Mitchell (Eds.), *Machine Learning*, Springer-Verlag, Berlin, 3-23.

CARNEVALLI, P., L. COLETTI, AND S. PATARNELLO [1985], Image processing by simulated annealing, *IBM Journ. of Research and Development* **29**, 569-579.

CASOTTO, A., F. ROMEO, AND A.L. SANGIOVANNI-VINCENTELLI [1987], A parallel simulated annealing algorithm for the placement of macro-cells, *IEEE Trans. on Computer-Aided Design* **6**, 838-847.

CASOTTO, A., AND A.L. SANGIOVANNI-VINCENTELLI [1987], Placement of standard cells using simulated annealing on the connection machine, *Proc. IEEE Int. Conf. on Computer-Aided Design*, Santa Clara, 350-353.

CATTHOOR, F., H. DEMAN, AND J. VANDERWALLE [1988], SAMURAI: a general and efficient simulated-annealing schedule with fully adaptive annealing parameters, *Integration* **6**, 147-178.

ČERNY, V. [1985], Thermodynamical approach to the traveling salesman problem: an efficient simulation algorithm, *Journ. of Optimization Theory and Applications* **45**, 41-51.

CHAMS, M., A. HERTZ, AND D. DE WERRA [1987], Some experiments with simulated annealing for colouring graphs, *European Journ. of Operational Research* **32**, 260-266.

CHEN, W.-M., Y.-X. WONG, AND X. PING [1987], Flow-shop scheduling by the knowledge of statistical mechanics and annealing, *Proc. 26th IEEE Conf. on Decision and Control* **1**, Los Angeles, 642-643.

COHEN, P.R., AND E.A. FEIGENBAUM [1982], *The Handbook of Artificial Intelligence* **3**, Pitman Books, London.

COLLINS, N.E., R.W. EGLESE, AND B.L. GOLDEN [1987], Simulated annealing - an annotated bibliography, Cambridge University, preprint.

Computer [1988], **21**.

Communications of the ACM [1988], **31**.

CONNORS, D.P., AND P.R. KUMAR [1987], Simulated annealing and balance of recurrence order in time-inhomogeneous Markov chains, *Proc. 26th IEEE Conf. on Decision and Control* **3**, Los Angeles, 2261-2263.

COOK, S.A. [1971], The complexity of theorem procedures, *Proc. 3rd ACM Symposium on the Theory of Computing*, 151-158.

COOK, S.A. [1972], An overview of computational complexity, *Communications of the ACM* **26**, 400-408.

CORANA, A., M. MARCHESI, C. MARTINI, AND S. RIDELLA [1987], Minimizing multimodal functions of continuous variables with the "Simulated Annealing" algorithm, *ACM Trans. on Mathimatical Software* **13**, 262-280.

DAHL, E.D. [1987], Neural network algorithm for an NP-complete problem: map and graph coloring, *Proc. Int. Conf. on Neural Networks* **3**, San Diego, 113-120.

DANTZIG, G.B. [1963], *Linear Programming and Extensions*, Princeton University Press, Princeton (NJ).

DAREMA-ROGERS, F., S. KIRKPATRICK, AND V.A. NORTON [1987], Parallel algorithms for chip placement by simulated annealing, *IBM Journ. of Research and Development* **31**, 391-402.

DEMPSTER, A.P., N.M. LAIRD, AND D.B. RUBIN [1977], Maximum likelihood from incomplete data via the EM algorithm, *Journ. of the Royal Statistical Society* **339**, 1-38.

DENKER, J.S. (ED.) [1986], *Neural Networks for Computing*, American Institute of Physics Conf. Proc. **151**, Snowbird (UT).

DENNIS, J., AND E. VAN HORN [1966], Programming semantics for multiprogrammed computations, *Communications of the ACM* **9**, 143-154.

DERRIDA, B., E. GARDNER, AND A. ZIPPELIUS [1987], An exactly solvable asymmetric neural network model, *Europhysics Letters* **4**, 167-173.

DERTHICK, M. [1984], Variations on the Boltzmann machine learning algorithm, Carnegie-Mellon University, *Technical Report CMU-CS-84-120*.

DEVADAS, S., AND A.R. NEWTON [1986], Topological optimization of multiple level array logic: on uni and multi-processors, *Proc. IEEE Int. Conf. on Computer-Aided Design*, Santa Clara, 38-41.

DEVIJVER, P.A., AND J. KITTLER (EDS.) [1987], *Pattern Recognition Theory and Applications*, NASI Series on Computer and Systems Sciences **30**, Springer-Verlag, Berlin.

DISTANTE, F., AND V. PIURI [1986], Optimum behavioral test procedure for VLSI devices: a simulated annealing approach, *Proc. IEEE Int. Conf. on Computer Design*, Port Chester, 31-35.

DOLEV, D., N.A. LYNCH, S.S. PINTER, E.W. STARK AND W.E. WEIHL [1986], Reaching approximate agreement in the presence of faults, Programming semantics for multiprogrammed computations, *Journ. of the ACM* **33**, 499-516.

DRESS A., AND M. KRUGER [1987], Parsimonious phylogenetic trees in metric spaces and simulated annealing, *Advances in Applied Mathematics* **8**, 8-37.

EGLESE, R.W., AND G.K. RAND [1987], Conference seminar timetabling, *Journ. of the Operational Research Society* **38**, 591-598.

EL GAMAL, A., L.A. HEMACHANDRA, I. SHPERLING, AND V.K. WEI [1987], Using simulated annealing to design good codes, *IEEE Trans. on Information Theory* **33**, 116-123.

FAHLMAN, S.E., AND G.E. HINTON [1987], Connectionist architectures for artificial intelligence, *Computer* **20**, 100-109.

FAHLMAN, S.E., G.E. HINTON AND T.J. SEJNOWSKI [1983], Massively parallel architectures for AI: NETL, Thistle and Boltzmann machines, *Proc. National Conf. on Artificial Intelligence*, AAAI-83, USA, 109-113.

FAIGLE, U., AND R. SCHRADER [1988], On the convergence of stationary distributions in simulated annealing algorithms, *Information Processing Letters* **27**, 189-194.

FARHAT, N.H. [1987], Optoelectronic analogs of self-programming neural nets: architecture and methodologies for implementing fast stochastic learning by simulated annealing, *Applied Optics* **26**, 5093-5103.

FELDMAN, J.A. [1981], A connectionist model of visual memory, in: G.E. Hinton and J.A. Anderson (Eds.), *Parallel Models of Associative Memory*, Erlbaum, Hillsdale (NJ), 49-81.

FELDMAN, J.A., AND D.H. BALLARD [1982], Connectionist models and their properties, *Cognitive Science* **6**, 205-254.

FELLER, W. [1950], *An Introduction to Probability Theory and Its Applications* **1**, Wiley, New York.

FELTEN, E., S. KARLIN, AND S.W. OTTO [1985], The traveling salesman problem on a hypercubic, MIMD computer, *Proc. 1985 Int. Conf. on Parallel Processing*, St. Charles, 6-10.

FLEISHER, H., J. GIRALDI, D.B. MARTIN, R.L. PHOENIX, AND M.A. TAVEL [1985], Simulated annealing as a tool for logic optimization in a CAD environment, *Proc. IEEE Int. Conf. on Computer-Aided Design*, Santa Clara, 203-205.

FLYNN, M.J. [1966], Very high-speed computing systems, *Proceedings IEEE* **54**, 1901-1909.

FORD, L.R. JR., AND D.R. FULKERSON [1962], *Flows in Networks*, Princeton University Press, Princeton (NJ).

FOULDS, L.R. [1983], Techniques for facilities layout: deciding which pairs of activities should be adjacent, *Management Science* **29**, 1426-1444.

FU, Y., AND P.W. ANDERSON [1986], Application of statistical mechanics to NP-complete problems in combinatorial optimisation, *Journ. of Physics A* **19**, 1605-1620.

FUKUSHIMA, K. [1975], Cognitron: A self-organizing multilayered neural network, *Biological Cybernetics* **20**, 121-136.

FUKUSHIMA, K. [1980], Neocognitron: A self-organizing neural network model for a mechanism of pattern recognition unaffected by shift in position, *Biological Cybernetics* **36**, 193-202.

GAREY, M.R., AND D.S. JOHNSON [1979], *Computers and Intractability: A Guide to the Theory of NP-Completeness*, W.H. Freeman and Co., San Francisco.

GARFINKEL, R.S. [1985], Motivation and modeling, in: E.L. Lawler, J.K. Lenstra, A.H.G. Rinnooy Kan, and D.B. Shmoys (Eds.), *The Traveling Salesman Problem: A Guided Tour of Combinatorial Optimization*, Wiley, Chichester, 17-36.

GELFAND, S.B., AND S.K. MITTER [1985], Analysis of simulated annealing for optimization, *Proc. 24th IEEE Conf. on Decision and Control*, Ft. Lauderdale, 779-786.

GEMAN, D., AND S. GEMAN [1986], Bayesian image analysis, in: E. Bienenstock, F. Fogelman Soulie and G. Weisbuch (Eds.), *Disordered Systems and Biological Organization*, NASI Series on Computer and Systems Sciences **20**, Springer-Verlag, Berlin, 301-319.

GEMAN, S. [1987], Stochastic relaxation methods for image restoration and expert systems, in: D.B. Cooper, R.L. Launer and D.E. McClure (Eds.), *Automated Image Analysis: Theory and Experiments*, Academic Press, New York, in press.

GEMAN, S., AND D. GEMAN [1984], Stochastic relaxation, Gibbs distributions, and the Bayesian restoration of images, *IEEE Proc. Pattern Analysis and Machine Intelligence* **6**, 721-741.

GEMAN, S., AND C.-R. HWANG [1986], Diffusions for global optimization, *SIAM Journ. of Control and Optimization* **24**, 1031-1043.

GENESERETH, M.R., M.L. GINSBERG, AND J.S. ROSENSCHEIN [1986], Cooperation without communication, *Proc. 5th National Conf. on Artificial Intelligence* **1**, AAAI-86, USA, 551-557.

GIBBS, J.W. [1902], *Elementary Principles in Statistical Mechanics*, Yale University Press, New Haven.

GIDAS, B. [1985a], Nonstationary Markov chains and convergence of the annealing algorithm, *Journ. of Statistical Physics* **39**, 73-131.

GIDAS, B. [1985b], Global optimization via the Langevin equation, *Proc. 24th IEEE Conf. on Decision and Control*, Ft. Lauderdale, 774-778.

GOLDEN, B.L., AND C.C. SKISCIM [1986], Using simulated annealing to solve routing and location problems, *Naval Logistics Research Quarterly* **33**, 261-279.

GOLDSTEIN, L., AND M.S. WATERMAN [1987], Mapping DNA by stochastic relaxation, *Advances in Applied Mathematics* **8**, 194-207.

GONSALVES, G. [1986], Logic synthesis using simulated annealing, *Proc. IEEE Int. Conf. on Computer Design*, Port Chester, 561-564.

GRAF, H.P., L.D. JACKEL, R.E. HOWARD, B. STAUGHN, J.S. DENKER, W. HUBBARD, D.M. TENNANT, AND D. SCHWARTZ [1986], VLSI implementation of a neural network memory with several hundreds of neurons, in: J.S. Denker (Ed.), *Neural Networks for Computing*, AIP Conf. Proc. **151**, Snowbird (UT), 182-187.

GREENE, J.W., AND K.J. SUPOWIT [1986], Simulated annealing without rejected moves, *IEEE Trans. on Computer-Aided Design* **5**, 221-228.

GROSSBERG, S. [1982], *Studies of Mind and Brain: Neural Principles of Learning, Perception, Development, Cognition and Motor Control*, Reidel, Boston.

GROSSBERG, S. (ED.) [1988a], *Neural Networks and Natural Intelligence*, MIT Press, Cambridge (MA).

GROSSBERG, S. [1988b], Nonlinear neural networks: principles, mechanisms, and architectures, *Neural Networks* **1**, 17-61.

GRÖTSCHEL, M. [1977], *Polyedrische Charakterisierungen Kombinatorischer Optimierungsprobleme* (in German), Hain, Meisenheim am Glan.

GRÖTSCHEL, M. [1984], Polyedrische Kombinatorik und Schnittebenenverfahren (in German), Universität Augsburg, preprint, no. 38.

GROVER, L.K. [1986], A new simulated annealing algorithm for standard cell placement, *Proc. IEEE Int. Conf. on Computer-Aided Design*, Santa Clara, 378-380.

GULATI, S., S.S. IYENGAR, N. TOOMARIAN, V. PROTOPOPESCU, AND J. BAHREN [1987], Nonlinear neural networks for deterministic scheduling, *Proc. Int. Conf. on Neural Networks* **4**, San Diego, 745-752.

GUTZMANN, K.M. [1987], Combinatorial optimization using a continuous state Boltzmann machine, *Proc. Int. Conf. on Neural Networks* **3**, San Diego, 721-734.

HAJEK, B. [1985], A tutorial survey of the theory and applications of simulated annealing, *Proc. 24th IEEE Conf. on Decision and Control*, Ft. Lauderdale, 755-760.

HAJEK, B. [1988], Cooling schedules for optimal annealing, *Mathematics of Operations Research* **13**, 311-329.

HARARY, F. [1972], *Graph Theory*, 3rd ed., Addison-Wesley, Reading (MA).

HEBB, D.O. [1949], *The Organisation of Behaviour*, Wiley, New York.

HECHT-NIELSEN, R. [1987], Performance of optical, electro-optical, and electronic neurocomputers, in: H. Szu (Ed.), *Optical and Hybrid Computing*, SPIE **634**, Bellingham (WA), 277-306.

HEMMEN, J.L. VAN [1982], Classical spin-glass model, *Physics Review Letters* **49**, 409-412.

HILLIS, W.D. [1985], *The Connection Machine*, MIT Press, Cambridge (MA).

HINTON, G.E. [1986], Learning distributed representations of concepts, *Proc. 8th Annual Conf. of the Cognitive Science Society*, Amherst (MA).

HINTON, G.E. [1987], Connectionist learning procedures, Carnegie-Mellon University, *Technical Report CMU-CS-87-115*.

HINTON, G.E., AND J.A. ANDERSON (EDS.) [1981], *Parallel Models of Associative Memory*, Erlbaum, Hillsdale (NJ).

HINTON, G.E., AND T.J. SEJNOWSKI [1983], Optimal perceptual inference, *Proc. IEEE Conf. on Computer Vision and Pattern Recognition*, Washington DC, 448-453.

HINTON, G.E., AND T.J. SEJNOWSKI [1986], Learning and relearning in Boltzmann machines, in: D.E. Rumelhart, J.L. McClelland and the PDP Research Group (Eds.), *Parallel Distributed Processing: Explorations in the Microstructure of Cognition* **1**, Bradford Books, Cambridge (MA), 282-317.

HINTON, G.E., T.J. SEJNOWSKI AND D.H. ACKLEY [1984], Boltzmann machines: constraint satisfaction networks that learn, Carnegie-Mellon University, *Technical Report CMU-CS-84-119*.

HOLLEY, R., AND D. STROOCK [1988], Simulated annealing via Sobolev inequalities, *Communications in Mathematical Physics* **115**, 553-569.

HOPFIELD, J.J. [1982], Neural networks and physical systems with emergent collective computational abilities, *Proc. National Academy of Sciences of the USA* **79**, 2554-2558.

HOPFIELD, J.J. [1984], Neurons with graded response have collective computational properties like those of two-state neurons, *Proc. National Academy of Sciences of the USA* **81**, 3088-3092.

HOPFIELD, J.J., AND D.W. TANK [1985], Neural computation of decisions in optimization problems, *Biological Cybernetics* **52**, 141-152.

HOPFIELD, J.J., AND D.W. TANK [1986], Computing with neural circuits: a model, *Science* **233**, 625-633.

HU, T.C. [1970], *Integer Programming and Network Flows*, Addison-Wesley, Reading (MA).

IOSUPOVICI, A., C. KING, AND M. BREUER [1983], A module interchange placement machine, *Proc. IEEE 20th Design Automation Conf.*, Miami Beach, 171-174.

ISAACSON, D., AND R. MADSEN [1976], *Markov Chains*, Wiley, New York.

JEPSEN, D.W., AND C.D. GELATT JR. [1983], Macro placement by Monte Carlo annealing, *Proc. IEEE Int. Conf. on Computer Design*, Port Chester, 495-498.

JOHNSON, D.S., C.R. ARAGON, L.A. MCGEOCH, AND C. SCHEVON [1987], Optimization by simulated annealing: an experimental evaluation, Part I, AT&T Bell Laboratories, Murray Hill (NJ), preprint.

JOHNSON, D.S., C.R. ARAGON, L.A. MCGEOCH, AND C. SCHEVON [1988], Optimization by simulated annealing: an experimental evaluation, Part II, AT&T Bell Laboratories, Murray Hill (NJ), preprint.

JOHNSON, D.S., C.H. PAPADIMITRIOU, AND M. YANNAKAKIS [1985], How easy is local search?, *Proc. Annual Symposium on Foundations of Computer Science*, Los Angeles, 39-42.

JONKER, R., AND T. VOLGENANT [1984], Nonoptimal edges for the travelling salesman problem, *Operations Research* **32**, 837-846.

KAMGAR-PARSI, B., AND B. KAMGAR-PARSI [1987], An efficient model of neural networks for optimization, *Proc. Int. Conf. on Neural Networks* **3**, San Diego, 785-790.

KARG, R.L., AND G.L. THOMPSON [1964], A heuristic approach to solving travelling salesman problems, *Management Science* **10**, 225-247.

KARP, R.M. [1972], Reducibility among combinatorial problems, in: R.E. Miller and J.W. Thatcher (Eds.), *Complexity of Computer Computations*, Plenum Press, New York, 85-103.

KERN, W. [1986a], A probabilistic analysis of the switching algorithm for the Euclidean TSP, Universität zu Köln, Köln, *Technical Report 86.28*.

KERN, W. [1986b], On the depth of combinatorial optimization problems, Universität zu Köln, Köln, *Technical Report 86.33*.

KERNIGHAN, B.W., AND S. LIN [1970], An efficient heuristic procedure for partitioning graphs, *Bell System Technical Journ.* **49**, 291-307.

KHACHATURYAN, A.G. [1986], Statistical mechanics approach in minimizing a multivariable function, *Journ. of Mathematical Physics* **27**, 1834-1838.

KIRKPATRICK, S. [1984], Optimization by simulated annealing: quantitative studies, *Journ. of Statistical Physics* **34**, 975-986.

KIRKPATRICK, S., C.D. GELATT JR., AND M.P. VECCHI [1982], Optimization by simulated annealing, *IBM Research Report RC 9355*.

KIRKPATRICK, S., C.D. GELATT JR., AND M.P. VECCHI [1983], Optimization by simulated annealing, *Science* **220**, 671-680.

KIRKPATRICK, S., AND G. TOULOUSE [1985], Configuration space analysis of travelling salesman problems, *Journ. de Physique* **46**, 1277-1292.

KOHONEN, T. [1987], *Content-Addressable Memories*, 2nd ed., Springer-Verlag, Berlin.

KOHONEN, T. [1988], *Self-Organization and Associative Memory*, 2nd ed., Springer-Verlag, Berlin.

KOHONEN, T., K. TORKKOLA, M. SHOZAKAI, J. KANGAS, AND O. VENTÄ [1987], Microprocessor implementation of a large vocabulary speech recognizer and phonetic typewriter for Finnish and Japanese, in: J.A. Laver and M.A. Jack (Eds.), *Proc. European Conf. on Speech Technology*, Edinburgh, 377-380.

KORST, J.H.M., AND E.H.L. AARTS [1988], Combinatorial optimization on a Boltzmann machine, *Proc. European Seminar on Neural Computing*, London, IBC Tech. Serv., London.

KOU, L.T., L.J. STOCKMEYER, AND C.K. WONG [1978], Covering edges by cliques with regard to keyword conflicts and intersection graphs, *Communications of the ACM* **21**, 135-139.

KRAVITZ, S.A., AND R. RUTENBAR [1987], Placement by simulated annealing on a multiprocessor, *IEEE Trans. on Computer-Aided Design* **6**, 534-549.

KULLBACK, S. [1959], *Information Theory and Statistics*, Wiley, New York.

LAARHOVEN, P.J.M. VAN [1988], *Theoretical and Computational Aspects of Simulated Annealing*, Erasmus University Rotterdam, Ph.D. thesis.

LAARHOVEN, P.J.M. VAN, AND E.H.L. AARTS [1987], *Simulated Annealing: Theory and Applications*, Reidel, Dordrecht.

LAARHOVEN, P.J.M. VAN, E.H.L. AARTS, AND J.K. LENSTRA [1988], Solving job shop scheduling problems by simulated annealing, Philips Research Laboratories, Eindhoven, preprint.

LAARHOVEN, P.J.M. VAN, E.H.L. AARTS, J.H. VAN LINT, AND L.T. WILLE [1988], New upper bounds for the football pool problem for 6,7 and 8 matches, *Journ. of Combinatorial Theory A*, in press.

LAM, J., AND J.-M. DELOSME [1986], Logic minimization using simulated annealing, *Proc. IEEE Int. Conf. on Computer-Aided Design*, Santa Clara, 348-351.

LAWLER, E.L. [1976], *Combinatorial Optimization: Networks and Matroids*, Holt, Rinehart and Winston, New York.

LAWLER, E.L., J.K. LENSTRA, A.H.G. RINNOOY KAN, AND D.B. SHMOYS (EDS.) [1985], *The Traveling Salesman Problem: A Guided Tour of Combinatorial Optimization*, Wiley, Chichester.

LEEUWEN, J. VAN [1985], Parallel computers and algorithms, in: J. van Leeuwen and J.K. Lenstra (Eds.), Parallel computers and computations, *CWI Syllabus* **9**, 1-32.

LEONG, H.W. [1986], A new algorithm for gate matrix layout, *Proc. IEEE Int. Conf. on Computer-Aided Design*, Santa Clara, 316-319.

LEONG, H.W., AND C.L. LIU [1985], Permutation channel routing, *Proc. IEEE Int. Conf. on Computer Design*, Port Chester, 579-584.

LEONG, H.W., D.F. WONG AND C.L. LIU [1985], A simulated-annealing channel router, *Proc. IEEE Int. Conf. on Computer-Aided Design*, Santa Clara, 226-229.

LEVIN, L.A. [1973], Universal sorting problems, *Problemy Peredaci Informacii* **9**, 115-116 (in Russian), English translation in: *Problems of Information Transmission* **9**, 265-266.

LEVY, B.C., AND M.B. ADAMS [1987], Global optimization with stochastic networks, *Proc. Int. Conf. on Neural Networks* **3**, San Diego, 681-690.

LIGTHART, M.M., E.H.L. AARTS, AND F.P.M. BEENKER [1986], A design-for-testability of PLA's using statistical cooling, *Proc. 23rd Design Automation Conf.*, Las Vegas, 339-345.

LIN, S. [1965], Computer solutions of the traveling salesman problem, *Bell System Technical Journ.* **44**, 2245-2269.

LIN, S., AND B.W. KERNIGHAN [1973], An effective heuristic algorithm for the traveling salesman problem, *Operations Research* **21**, 498-516.

LITTLE, W. A. [1974], The existence of persistent states in the brain, *Mathematical Biosciences* **19**, 101-120.

LITTLE, W.A., AND G.L. SHAW [1978], Analytic study of the memory storage capability of a neural network, *Mathematical Biosciences* **39**, 281-290.

LUENBERGER, D.G. [1973], *Introduction to Linear and Nonlinear Programming*, Addison-Wesley, Reading (MA).

LUNDY, M. [1985], Applications of the annealing algorithm to combinatorial problems in statistics, *Biometrika* **72**, 191-198.

LUNDY, M., AND A. MEES [1986], Convergence of an annealing algorithm, *Mathematical Programming* **34**, 111-124.

LUTTON, J.-L., AND E. BONOMI [1986], Simulated annealing algorithm for the minimum weighted perfect Euclidean matching problem, *Recherche Opérationnelle* **20**, 177-197.

LYBERATOS, A., P. WOHLFARTH, AND R.W. CHANTRELL [1985], Simulated annealing: an application in fine particle magnetism, *IEEE Trans. on Magnetics* **21**, 1277-1282.

MCCULLOCH, W.S., AND W. PITTS [1943], A logical calculus of the ideas imminent in nervous activity, *Bulletin of Mathematics and Biophysics* **5**, 115-133.

MEAD, C.A. [1988], *Analog VLSI and Neural Systems*, Addison-Wesley, Reading (MA).

METROPOLIS, N., A. ROSENBLUTH, M. ROSENBLUTH, A. TELLER, AND E. TELLER [1953], Equation of state calculations by fast computing machines, *Journ. of Chemical Physics* **21**, 1087-1092.

MÉZARD, M. [1987], Spin glasses and optimization, in: J.L. van Hemmen and I. Morgenstern (Eds.), *Lecture Notes in Physics* **275**, Springer-Verlag, Berlin, 354-372.

MÉZARD, M., AND G. PARISI [1985], Replicas and optimization, *Journ. de Physique Lettres* **46**,771-778. Spin glasses and optimization, in: J.L. van Hemmen and I. Morgenstern (Eds.), *Lecture Notes in Physics* **275**, Springer-Verlag, Berlin, 354-372.

MICHALSKI, R.S., J.G. CARBONELL, AND T.M. MITCHELL (EDS.) [1983], *Machine Learning: An Artificial Intelligence Approach*, Springer-Verlag, Berlin.

MILLER, C.E., AND J.W. THATCHER (EDS.) [1972], *Complexity of Computer Computations*, Plenum Press, New York.

MINSKY, M., AND S. PAPERT [1969], *Perceptrons*, MIT Press, Cambridge (MA).

MITRA, D., ROMEO, F., AND A.L. SANGIOVANNI-VINCENTELLI [1986], Convergence and finite-time behavior of simulated annealing, *Advances in Applied Probability* **18**, 747-771.

MORGENSTERN, C.A., AND H.D. SHAPIRO [1986], Chromatic number approximation using simulated annealing, Department of Computer Science, The University of New Mexico, Albuquerque, *Technical Report CS86-1*.

MORGENSTERN, I. [1987], Spin glasses, optimization and neural networks, in: J.L. van Hemmen and I. Morgenstern (Eds.), *Lecture Notes in Physics* **275**, Springer-Verlag, Berlin, 399-427.

MOUSSOURIS, J., [1974], Gibbs and Markov random systems with constraints, *Journ. of Statistical Physics* **10**, 11-33.

MURRAY, D.W., AND B.F. BUXTON [1987], Scene segmentation from visual-motion using global optimization, *IEEE Trans. on Pattern Analysis and Machine Intelligence* **9**, 220-226.

NAHAR, S., S. SAHNI, AND E. SHRAGOWITZ [1985], Experiments with simulated annealing, *Proc. 22nd Design Automation Conf.*, Las Vegas, 748-752.

NAHAR, S., S. SAHNI, AND E. SHRAGOWITZ [1986], Simulated annealing and combinatorial optimization, *Proc. 23rd Design Automation Conf.*, Las Vegas, 293-299.

Neural Networks [1988], **1**.

NICHOLSON, D.M., A. CHOWDHARY, AND L. SCHWARTZ [1984], Monte Carlo optimization of pair distribution functions - application to the electronic structure of disordered metals, *Physical Review B* **19**, 1633-1637.

OTTEN, R.H.J.M., AND L.P.P.P. VAN GINNEKEN [1984], Floorplan design using simulated annealing, *Proc. IEEE Int. Conf. on Computer-Aided Design*, Santa Clara, 96-98.

PALMER, E. [1985], *Graphical Evolution*, Wiley, New York.

PAPADIMITRIOU, C.H., AND K. STEIGLITZ [1982], *Combinatorial Optimization: Algorithms and Complexity*, Prentice-Hall, New York.

PAXMAN, R.G., W.E. SMITH, AND H.H. BARRETT [1984], Two algorithms for use with an orthogonal-view coded-aperture system, *Journ. of Nuclear Medicine* **25**, 700-705.

PERETTO, P. [1984], Collective properties of neural networks: a statistical physics approach, *Biological Cybernetics* **50**, 51-62.

PERSONNAZ, L., I. GUYON, AND G. DREYFUS [1985], Information storage and retrieval in spin-glass like neural networks, *Journ. de Physique Lettres* **46**, 359-365.

PERUSCH, M. [1987], Simulated annealing applied to a single machine scheduling problem with sequence-dependent setup times and due dates, *Belgian Journ. of Operational Research, Statistics, and Computer Science* **27**.

POSCH, T.E. [1968], Models of the generation and processing of signals by nerve cells: a categorical indexed abridged bibliography, University of Southern California, *Research Report USCEE 290*.

PRAGER, R.G., T.D. HARRISON, AND F. FALLSIDE [1986], Boltzmann machines for speech recognition, *Computer Speech and Language* **1**, 3-27.

Proceedings of the Int. Conf. on Neural Networks [1987], San Diego.

RANDELMAN, R.E., AND G.S. GREST [1986], N-city traveling salesman problem: optimization by simulated annealing, *Journ. of Statistical Physics* **45**, 885-890.

RIPLEY, B.D. [1986], Statistics, images and pattern recognition, *Canadian Journ. of Statistics* **14**, 83-102.

ROMEO, F., AND A.L. SANGIOVANNI-VINCENTELLI [1985], Probabilistic hill climbing algorithms: properties and applications, *Proc. Chapel Hill Conf. on VLSI*, Chapel Hill (NC), 393-417.

ROMEO, F., A.L. SANGIOVANNI-VINCENTELLI, AND C. SECHEN [1984], Research on simulated annealing at Berkeley, *Proc. IEEE Int. Conf. on Computer Design*, Port Chester, 652-657.

ROSE, J.S., D.R. BLYTHE, W.M. SNELGROVE, AND Z.G. VRANESIC [1986], Fast, high quality VLSI placement on an MIMD multiprocessor, *Proc. IEEE Int. Conf. on Computer-Aided Design*, Santa Clara, 42-45.

ROSENBLATT, F. [1962], *Principles of Neurodynamics*, Spartan Books, Washington DC.

ROSSIER, Y., M. TROYON, AND T.M. LIEBLING [1986], Probabilistic exchange algorithms and Euclidean travelling salesman problems, *OR Spectrum* **8**, 151-164.

ROTHMAN, D.H. [1985], Nonlinear inversion, statistical mechanics, and residual statics estimation, *Geophysics* **50**, 2784-2796.

ROWEN, C., AND J.L. HENNESSY [1985], SWAMI: a flexible logic implementation system, *Proc. 22nd Design Automation Conf.*, Las Vegas, 169-175.

RUMELHART, D.E., G.E. HINTON, AND R.J. WILLIAMS [1986], Learning internal representations by error propagation, in: D.E. Rumelhart, J.L. McClelland, and the PDP Research Group (Eds.), *Parallel Distributed Processing: Explorations in the Microstructure of Cognition* **1**, Bradford Books, Cambridge (MA), 318-362.

RUMELHART, D.E., J.L. MCCLELLAND, AND THE PDP RESEARCH GROUP (EDS.) [1986], *Parallel Distributed Processing: Explorations in the Microstructure of Cognition*, Bradford Books, Cambridge (MA).

SAGE, J.P., K. THOMPSON, AND R.S. WITHERS [1986], An artificial neural network integrated circuit based on NMOS/CCD principles, in: J.S. Denker (Ed.), *Neural Networks for Computing*, AIP Conf. Proc. **151**, Snowbird (UT), 381-385.

SASAKI, G.H., AND B. HAJEK [1988], The time complexity of maximum matching by simulated annealing, *Journ. of the ACM* **35**, 387-403.

SCHRIJVER, A. [1986], *Theory of Linear and Integer Programming*, Wiley, Chichester.

SCHWARTZ, J.T. [1980], Ultracomputers, *ACM Trans. on Programming Languages and Systems* **2**, 484-521.

SECHEN, C., AND A.L. SANGIOVANNI-VINCENTELLI [1985], The Timber-Wolf placement and routing package, *IEEE Journ. on Solid State Circuits* **30**, 510-522.

SEJNOWSKI, T.J. [1986], Higher order Boltzmann machines, in: J.S. Denker (Ed.), *Neural Networks for Computing*, AIP Conf. Proc. **151**, Snowbird (UT), 398-403.

SEJNOWSKI, T.J., AND G.E. HINTON [1985], Separating figures from ground with a Boltzmann machine, in: M.A. Arbib, and A.R. Hanson (Eds.), *Vision, Brain and Cooperative Computation*, MIT Press, Cambridge (MA), 703-724.

SEJNOWSKI, T.J., P.K. KIENKER, AND G.E. HINTON [1986], Learning symmetry groups with hidden units: beyond the perceptron, *Physica D* **22**, 260-275.

SEJNOWSKI, T.J., AND C.R. ROSENBERG [1987], Parallel networks that learn to pronounce English text, *Complex Systems* **1**, 145-168.

SEMENOVSKAYA, S.V., K.A. KHACHATURYAN, AND A.G. KHACHATU-RYAN [1985], Statistical mechanics approach to the structure determination of a crystal, *Acta Crystallographica A* **41**, 268-273.

SENETA, E. [1981], *Non-negative Matrices and Markov Chains*, 2nd ed., Springer-Verlag, New York.

SHANNON, C.E. [1948], A mathematical theory of communication, *Bell System Technical Journ.* **27**, 379-623.

SHARPE, R., AND B.S. MARKSJO [1985], Facility layout optimization using the Metropolis algorithm, *Environment and Planning B* **12**, 443-453.

SHARPE, R., B.S. MARKSJO, J.R. MITCHELL, AND J.R. CRAWFORD [1985], An interactive model for the layout of buildings, *Applied Mathematics and Modelling* **9**, 207-214.

SHEILD, J. [1987], Partitioning concurrent VLSI simulation programs onto a multi-processor by simulated annealing, *IEE Proc. on Computers and Digital Techniques* **134**, 24-30.

SIARRY, P., L. BERGONZI, AND G. DREYFUS [1987], Thermodynamic optimization of block placement, *IEEE Trans. on Computer-Aided Design* **6**, 211-221.

SIMMONS, G.F. [1963], *Introduction to Topology and Modern Analysis*, McGraw-Hill, Kogakusha, Tokyo.

SIMONNARD, M. [1966], *Linear Programming*, Prentice Hall, Englewood Cliffs (NJ).

SKISCIM, C.C., AND B.L. GOLDEN [1983], Optimization by simulated annealing: a preliminary computational study for the TSP, *Proc. Winter Simulation Conf.* **2**, Arlington (VA), 523-535.

SMITH, W.E., H.H. BARRETT, AND R.G. PAXMAN [1983], Reconstruction of objects from coded images by simulated annealing, *Optics Letters* **8**, 199-201.

SONTAG, E.D., AND H.J. SUSSMANN [1986], Image restoration and segmentation using the annealing algorithm, *Proc. 24th Conf. on Decision and Control*, Ft. Lauderdale, 768-773.

SOUKUP, J. [1981], Circuit layout, *Proc. IEEE* **69**, 1281-1304.

SOURLAS, N. [1986], Statistical mechanics and the travelling salesman problem, *Europhysics Letters* **2**, 919-923.

SPIRA, P., AND C. HAGE [1985], Hardware acceleration of gate array layout, *Proc. 22nd Design Automation Conf.*, Las Vegas, 359-366.

STORER, J.A., J. BECKER, AND A.J. NICAS [1985], Uniform circuit placement, in: P. Bertolazzi and F. Luccio (Eds.), *Proc. Int. Workshop on Parallel Computing and VLSI*, Amalfi, Elsevier Science Publishers, Amsterdam, 255-273.

SZU, H., AND R. HARTLEY [1987], Fast simulated annealing, *Physics Letters A* **122**, 157-162.

TAGLIARINI, G.E., AND E.W. PAGE [1987], Solving constraint satisfaction problems with neural networks, *Proc. Int. Conf. on Neural Networks* **3**, San Diego, 741-748.

TANK, D.W., AND J.J. HOPFIELD [1987], Neural computation by concentrating information in time, *Proc. National Academy of Sciences of the USA* **84**, 1896-1900.

TELLEY, H., T.M. LIEBLING, AND A. MOCELLIN [1987], Reconstruction of polycrystalline structures: a new application of combinatorial optimization, *Computing* **38**, 1-11.

THOMPSON, R.F. [1986], The neurobiology of learning and memory, *Science* **233**, 941-947.

TICKNOR, A.J., AND H.H. BARRETT [1987], Optical implementations in Boltzmann machines, *Optical Engineering* **26**, 16-21.

TODA, M., R. KUBO, AND N. SAITÔ [1983], *Statistical Physics*, Springer-Verlag, Berlin.

TOULOUSE, G., S. DEHAENE, J. CHANGEUX [1986], Spin-glass models of neural networks, *Proc. National Academy of Sciences of the USA* **83**, 1695-1698.

TOURETZKY, D.S. [1986], Representing and transforming recursive objects in a neural network, or 'trees *do* grow on Boltzmann machines', *Proc. 1986 IEEE Conf. on Systems, Man, and Cybernetics*, Atlanta, 12-16.

TREHERN, J.F., M.A. JACK, AND J. LAVER [1986], Speech processing with a Boltzmann machine, *Proc. IEEE Int. Conf. on Acoustics, Speech, and Signal Processing*, 721-724.

TRELEAVEN, P.C. [1988], Parallel architectures for neurocomputers, *Proc. European Seminar on Neural Computing*, London, IBC Tech. Serv., London.

ULLMAN, J.D. [1984], *Computational Aspects of VLSI*, Computer Science Press, Rockville (MD).

VANDERBILT, D., AND S.G. LOUIE [1984], A Monte-Carlo simulated annealing approach to optimization over continuous variables, *Journ. of Computational Physics* **36**, 259-271.

VECCHI, M.P., AND S. KIRKPATRICK [1983], Global wiring by simulated annealing, *IEEE Transactions on Computer-Aided Design* **2**, 215-222.

WEBER M., AND T.M. LIEBLING [1986], Euclidean matching problems and the Metropolis algorithm, *Zeitschrift für Operations Research* **30**, 85-110.

WHITE, S.R. [1984], Concepts of scale in simulated annealing, *Proc. IEEE Int. Conf. on Computer Design*, Port Chester, 646-651.

WILHELM, M.R., AND T.L. WARD [1987], Solving quadratic assignment problems by simulated annealing, *IEE Trans. on Industrial Engineering Research and Development* **19**, 107-119.

WILLE, L.T. [1986a], A bibliography of papers on simulated annealing and related optimization techniques, SERC Daresbury Laboratory, Warrington, *Technical Report*.

WILLE, L.T. [1986b], Searching potential energy surfaces by simulated annealing, *Nature* **324**, 46-47.

WILLE, L.T. [1987], The football pool problem for 6 matches: a new upper bound obtained by simulated annealing, *Journ. of Combinatorial Theory A* **45**, 171-177.

WILLE, L.T., AND J. VENNIK [1985], Electrostatic energy minimization by simulated annealing, *Journ. of Physics A* **18**, 1983-1990.

WOLBERG, G., AND T. PAVLIDIS [1985], Restoration of binary images using stochastic relaxation with annealing, *Pattern Recognition Letters* **3**, 375-388.

WONG, D.F., H.W. LEONG, AND C.L. LIU [1986], Multiple PLA folding by the method of simulated annealing, *Proc. 1986 Custom IC Conf.*, Rochester, 351-355.

WONG, D.F., H.W. LEONG, AND C.L. LIU [1988], *Simulated Annealing for VLSI Design*, Kluwer, Boston.

WONG, D.F., AND C.L. LIU [1986], A new algorithm for floorplan design, *Proc. 23rd Design Automation Conf.*, Las Vegas, 101-107.

WOOTEN, F., K. WINER, AND D. WEAIRE [1985], Computer generation of structural models of amorphous Si and Ge, *Physics Review Letters* **54**, 1392-1395.

YOUNG, T.Y., AND T.W. CALVERT [1974], *Classification, Estimation, and Pattern Recognition*, Elsevier, New York.

List of Symbols

Most of the letters in the Roman and Greek alphabets have been used as symbols in this book, some even in more than one meaning. Those symbols which are used in one specific meaning throughout the book are listed below, separated in three groups: Roman, Greek and miscellaneous symbols.

A_{ij}	acceptance probability
$A_k(u, c)$	acceptance probability of unit u at a given value of c
c	control parameter
c_0	initial value of the control parameter
C	consensus function
C_{opt}	optimal consensus
$D(\mathbf{q}|\mathbf{q}')$	divergence measure of $\mathbf{q}$ with respect to $\mathbf{q}'$
f	cost function
f_{opt}	optimal cost
G_{ij}	generation probability
$G(u)$	probability of generating unit u
$k(u)$	state of unit u in configuration k
K	number of consecutive Markov chains in the stop criterion
L	Markov-chain length
$N_2(p, q)$	2-change mechanism
P_{ij}	transition probability
$p_{\{u,v\}}$	activation probability of connection $\{u, v\}$
$\mathbf{q}$	stationary distribution
$\mathbf{q}^*$	stationary distribution at $c = 0$
$R_k(u, c)$	rejection probability of unit u at a given value of c
$s_{\{u,v\}}$	connection strength of connection $\{u, v\}$
S	entropy
α	lowering factor of the control parameter

δ	distance parameter
$\Delta C_k(u)$	consensus difference of changing the state of unit u in configuration k
ε_s	stop parameter
Θ	neighbourhood size
σ	spreading
σ^2	variance
$\chi_{(A)}(a)$	characteristic function
χ	acceptance ratio
χ_0	initial acceptance ratio
$\mathcal{C}$	set of connections
$\mathcal{C}_u$	set of connections incident with unit u
$\mathbb{E}(x)$	expectation of x
$\mathcal{E}$	error
$\mathcal{H}$	internal configuration space
$\mathcal{I}$	input configuration space
$\mathcal{L}_A(k)$	projection of k onto subspace $\mathcal{A}$
$\mathcal{O}$	output configuration space
$\mathbb{P}\{A\vert B\}$	probability of event A given event B
$\mathcal{R}$	configuration space
$\mathcal{R}_i$	set of neighbouring configurations (neighbourhood)
$\mathcal{R}_{opt}$	set of optimal configurations
$\widehat{\mathcal{R}}$	set of locally-optimal configurations
$\mathcal{S}$	solution space
$\mathcal{S}_i$	set of neighbouring solutions (neighbourhood)
$\mathcal{S}_{opt}$	set of optimal solutions
$\widehat{\mathcal{S}}$	set of locally-optimal solutions
$\mathcal{U}$	set of units
$\mathcal{V}$	environmental configuration space
$\mathcal{V}'$	classification or learning set
$\langle x \rangle$	expected value of x
$\overline{x}$	average value of x

Author Index

261

Subject Index